Mastering PostGIS

Create, deliver, and consume spatial data using PostGIS

Dominik Mikiewicz
Michal Mackiewicz
Tomasz Nycz

BIRMINGHAM - MUMBAI

Mastering PostGIS

First published: May 2017

Production reference: 1260517

Published by Packt Publishing Ltd.
Livery Place
35 Livery Street
Birmingham
B3 2PB, UK.

ISBN 978-1-78439-164-5

www.packtpub.com

Credits

Authors
Dominik Mikiewicz
Michal Mackiewicz
Tomasz Nycz

Copy Editor
Safis Editing

Reviewers
Prashant Verma
Eric Pimpler

Project Coordinator
Nidhi Joshi

Commissioning Editor
Amey Varangaonkar

Proofreader
Safis Editing

Acquisition Editor
Vinay Argekar

Indexer
Mariammal Chettiyar

Content Development Editor
Mayur Pawanikar

Graphics
Tania Dutta

Technical Editor
Dinesh Chaudhary

Production Coordinator
Aparna Bhagat

About the Authors

Dominik Mikiewicz is a senior GIS consultant and the owner of one-person software shop Cartomatic. When not coding, he spends time with wife and kids, trying to make the little ones enjoy mountain trekking. He is also a long-distance cycling and running enthusiast.

Michal Mackiewicz has been working as a software engineer at GIS Support for five years. His main job is to orchestrate various open source geospatial components and creating application-specific GIS systems. PostgreSQL and PostGIS are among his favorite tools, and are used in almost every project. Apart from developing, he also runs PostGIS training courses. When not at work, he volunteers for OpenStreetMap and a local railway preservation society.

Tomasz Nycz is a geographer and cartographer. He initiated the implementation of GIS in the State Fire Service units in Poland. He works with recognized GIS companies in the emergency management industry. In practice, he uses QGIS and PostGIS. He has been an OpenStreetMap contributor for years. He also develops his scientific interests in the use of new technologies in geomorphology and remote sensing. He is also an avid drone pilot and mountain explorer.

About the Reviewers

Eric Pimpler is the founder and owner of GeoSpatial Training Services (geospatialtraining.com) and has over 20 years of experience in implementing and teaching GIS solutions using Esri, Google Earth/Maps, and open source technology. Currently, he focuses on ArcGIS scripting with Python and the development of custom ArcGIS Server web and mobile applications using JavaScript. He is the author of *Programming ArcGIS 10.1 with Python Cookbook*.

Eric has a bachelor's degree in Geography from Texas A&M University and a master's degree in Applied Geography with a concentration in GIS from Texas State University.

Prashant Verma started his IT carrier in 2011 as a Java developer at Ericsson working in the telecoms domain. After a couple of years of Java EE experience, he moved into the big data domain, and has worked on almost all the popular big data technologies, such as Hadoop, Spark, Flume, Mongo, and Cassandra. He has also played with Scala. Currently, he works with QA Infotech as lead data engineer, working on solving e-Learning problems using analytics and machine learning.
Prashant has also worked on *Apache Spark for Java Developers* as a technical reviewer.

I want to thank Packt Publishing for giving me the chance to review the book, as well as my employer and my family for their patience while I was busy working on this book.

www.PacktPub.com

For support files and downloads related to your book, please visit www.PacktPub.com.

Did you know that Packt offers eBook versions of every book published, with PDF and ePub files available? You can upgrade to the eBook version at www.PacktPub.com and as a print book customer, you are entitled to a discount on the eBook copy. Get in touch with us at service@packtpub.com for more details.

At www.PacktPub.com, you can also read a collection of free technical articles, sign up for a range of free newsletters and receive exclusive discounts and offers on Packt books and eBooks.

https://www.packtpub.com/mapt

Get the most in-demand software skills with Mapt. Mapt gives you full access to all Packt books and video courses, as well as industry-leading tools to help you plan your personal development and advance your career.

Why subscribe?

- Fully searchable across every book published by Packt
- Copy and paste, print, and bookmark content
- On demand and accessible via a web browser

Customer Feedback

Thanks for purchasing this Packt book. At Packt, quality is at the heart of our editorial process. To help us improve, please leave us an honest review on this book's Amazon page at `https://www.amazon.com/dp/1784391646`.

If you'd like to join our team of regular reviewers, you can e-mail us at `customerreviews@packtpub.com`. We award our regular reviewers with free eBooks and videos in exchange for their valuable feedback. Help us be relentless in improving our products!

Table of Contents

Preface

PostGIS is an open source extension of the PostgreSQL object-relational database system that allows GIS objects to be stored and allows querying for information and location services. The aim of this book is to help you master the functionalities offered by PostGIS, from data creation, analysis, and output to ETL and live edits.

The book begins with an overview of the key concepts related to spatial database systems and how it applies to spatial RMDS. You will learn to load different formats into your Postgres instance, investigate the spatial nature of your raster data, and finally export it using built-in functionalities or third-party tools for backup or representational purposes.

Through the course of this book, you will be presented with many examples on how to interact with the database using JavaScript and Node.js. Sample web-based applications interacting with backend PostGIS will also be presented throughout the book, so you can get comfortable with the modern ways of consuming and modifying your spatial data.

What this book covers

Chapter 1, *Importing Spatial Data*, will cover simple import procedures to import data to PgSQL/PostGIS.

Chapter 2, *Spatial Data Analysis*, looks at vector data analysis, and we'll find our way through a rich function set of PostGIS.

Chapter 3, *Data Processing - Vector Ops*, discusses the functions available for vector data processing.

Chapter 4, *Data Processing - Raster Ops*, discusses the functions available for raster data processing.

Chapter 5, *Exporting Spatial Data*, looks into exporting a dataset from PostGIS to other GIS formats.

Chapter 6, *ETL Using Node.js*, explains how to perform ETL ops using JavaScript in Node.js.

Chapter 7, *PostGIS – Creating Simple WebGIS Applications*, focuses on publishing PostGIS data with the usage of web platforms.

Chapter 8, *PostGIS Topology*, we will discusses different PostGIS Topology types and functions that are used to manage topological objects such as faces, edges, and nodes.

Chapter 9, *pgRouting*, explains the pgRouting extension and its implementations.

What you need for this book

This book will guide you through the installation of all the tools that you need to follow the examples.

Following is the list of software and the download link to work through this book:

- PostgreSQL 9.x (https://www.postgresql.org/download/)
- PostGIS 2.x (http://postgis.net/install/)
- QGIS 2.x (http://www.qgis.org/en/site/forusers/download.html)
- ogr2ogr / gdal (http://ogr2gui.ca/)
- Manifold 8 (http://manifold.net/updates/downloads.shtml)
- SQL SERVER 2016 (https://www.microsoft.com/en-us/sql-server/sql-server-downloads)
- pgAdmin 3 (Should be bundled with PostgreSQL, if not https://www.pgadmin.org/download/)
- OL3 (https://openlayers.org/download/)
- Leaflet (http://leafletjs.com/download.html)
- GeoServer 2.9 + with bundled jetty (http://geoserver.org/)
- ExtJs (https://www.sencha.com/products/evaluate/)
- Node.js (https://nodejs.org/en/download/)
- pgRouting (http://pgrouting.org/download.html)

Who this book is for

If you are a GIS developer or analyst who wants to master PostGIS to build efficient, scalable GIS applications, this book is for you. If you want to conduct advanced analysis of spatial data, this book will also help you. The book assumes that you have a working installation of PostGIS in place, and have working experience with PostgreSQL.

Conventions

In this book, you will find a number of text styles that distinguish between different kinds of information. Here are some examples of these styles and an explanation of their meaning.

Code words in text, database table names, folder names, filenames, file extensions, pathnames, dummy URLs, user input, and Twitter handles are shown as follows: "The next lines of code read the link and assign it to the `BeautifulSoup` function."

A block of code is set as follows:

```
drop table if exists data_import.osgb_addresses;
create table data_import.osgb_addresses(
   uprn bigint,
   os_address_toid varchar,
```

Any command-line input or output is written as follows:

```
mastering_postgis=# \copy data_import.earthquakes_csv from data\2.5_day.csv
with DELIMITER ',' CSV HEADER
```

New terms and **important words** are shown in bold. Words that you see on the screen, for example, in menus or dialog boxes, appear in the text like this: "In order to download new modules, we will go to **Files** | **Settings** | **Project Name** | **Project Interpreter**."

Warnings or important notes appear in a box like this.

Tips and tricks appear like this.

Reader feedback

Feedback from our readers is always welcome. Let us know what you think about this book-what you liked or disliked. Reader feedback is important for us as it helps us develop titles that you will really get the most out of.

To send us general feedback, simply e-mail `feedback@packtpub.com`, and mention the book's title in the subject of your message.

If there is a topic that you have expertise in and you are interested in either writing or contributing to a book, see our author guide at `www.packtpub.com/authors`.

Customer support

Now that you are the proud owner of a Packt book, we have a number of things to help you to get the most from your purchase.

Downloading the example code

You can download the example code files for this book from your account at `http://www.packtpub.com`. If you purchased this book elsewhere, you can visit `http://www.packtpub.com/support` and register to have the files e-mailed directly to you.

You can download the code files by following these steps:

1. Log in or register to our website using your e-mail address and password.
2. Hover the mouse pointer on the **SUPPORT** tab at the top.
3. Click on **Code Downloads & Errata**.
4. Enter the name of the book in the **Search** box.
5. Select the book for which you're looking to download the code files.
6. Choose from the drop-down menu where you purchased this book from.
7. Click on **Code Download**.

Once the file is downloaded, please make sure that you unzip or extract the folder using the latest version of:

- WinRAR / 7-Zip for Windows
- Zipeg / iZip / UnRarX for Mac
- 7-Zip / PeaZip for Linux

The code bundle for the book is also hosted on GitHub at `https://github.com/PacktPublishing/Mastering-Postgis`. We also have other code bundles from our rich catalog of books and videos available at `https://github.com/PacktPublishing/`. Check them out!

Downloading the color images of this book

We also provide you with a PDF file that has color images of the screenshots/diagrams used in this book. The color images will help you better understand the changes in the output. You can download this file from `https://www.packtpub.com/sites/default/files/down loads/MasteringPostgis_ColorImages.pdf`.

Errata

Although we have taken every care to ensure the accuracy of our content, mistakes do happen. If you find a mistake in one of our books-maybe a mistake in the text or the code-we would be grateful if you could report this to us. By doing so, you can save other readers from frustration and help us improve subsequent versions of this book. If you find any errata, please report them by visiting `http://www.packtpub.com/submit-errata`, selecting your book, clicking on the **Errata Submission Form** link, and entering the details of your errata. Once your errata are verified, your submission will be accepted and the errata will be uploaded to our website or added to any list of existing errata under the Errata section of that title.

To view the previously submitted errata, go to `https://www.packtpub.com/books/conten t/support` and enter the name of the book in the search field. The required information will appear under the **Errata** section.

Piracy

Piracy of copyrighted material on the Internet is an ongoing problem across all media. At Packt, we take the protection of our copyright and licenses very seriously. If you come across any illegal copies of our works in any form on the Internet, please provide us with the location address or website name immediately so that we can pursue a remedy.

Please contact us at `copyright@packtpub.com` with a link to the suspected pirated material.

We appreciate your help in protecting our authors and our ability to bring you valuable content.

Questions

If you have a problem with any aspect of this book, you can contact us at `questions@packtpub.com`, and we will do our best to address the problem.

1
Importing Spatial Data

Learning database tools means working with data, so we need to cover that aspect first. There are many ways of importing data to PgSQL/PostGIS; some are more database-specific, some are PostGIS-specific, and some use external tools. To complicate things a bit more, quite often real-world data import processes are wrapped into programs that perform different tasks and ops in order to maintain the data quality and integrity when importing it. The key though is that even very complex import tools usually use simpler procedures or commands in order to achieve their goals.

Such simple import procedures are described in this chapter. We specifically focus on:

- Importing flat data through both psql and pgAdmin and extracting spatial information from flat data
- Importing shape files using `shp2pgsql`
- Importing vector data using `ogr2ogr`
- Importing vector data using GIS clients
- Importing OpenStreetMap data
- Connecting to external data sources with data wrappers
- Loading rasters using `raster2pgsql`
- Importing data with pgrestore

Obtaining test data

Before we start importing, let's get some data examples in different formats, specifically these:

- Earthquake data in CSV and KML format (https://earthquake.usgs.gov/earthquakes/map/)
- UK Ordnance Survey sample data (https://www.ordnancesurvey.co.uk/business-and-government/licensing/sample-data/discover-data.html)
  - AddressBase in CSV and GML
  - Code-Point Polygons in SHP, TAB and MIF
  - Points of Interest in TXT format
- NaturalEarth (http://www.naturalearthdata.com/downloads/110m-physical-vectors/)
  - 110M coastlines
  - 110M land
  - 50M Gray Earth

You may either download the data using the links provided or find it in this chapter's resources.
The location you extract the data to is not important really, as you can later address it using either relative or absolute file paths.

Setting up the database

All the examples in this chapter use a database named `mastering_postgis`. This database has been created off the `postgis` template.

The PgSQL on my dev machine listens on port `5434`, which is not the default port for the Postgres database (default is `5432`); so when using a default DB setup, you may have to adjust some of the examples a bit.

If you need to change the port your db listens on, you should locate the db data directory, where you will find a `postgresql.conf` file. This is a text file, so you can edit it with an ordinary text editor.
In order to adjust the port, find a `port` configuration in the *Connections and Authentication* section.

Schemas are a great way of managing the data and splitting it into meaningful collections. In most scenarios, one will have some production data, archive data, incoming data, and so on sensibly kept in separate schemas. Using additional schemas will depend on your requirements, but we do encourage you to introduce using schemas into your daily practice if you do not yet do so. The following examples import the data into tables defined in the `data_import` schema.

Importing flat data

Loading flat data may seem to be a bit dull initially but it is important to understand that many popular and interesting datasets often contain the spatial information in very different formats, such as:

- Coordinates expressed in Lon/Lat or projected coordinates
- Encoded geometry, for example WKT, TopoJSON, GeoJSON
- Location in the form of an address
- Location in non-cartesian coordinates, for example start point, angle and direction
- While the earlier examples indicate the data would require further processing in order to extract the spatial content into a usable form, clearly ability to import flat datasets should not be underestimated

Flat data in our scenario is data with no explicitly expressed geometry - non-spatial format, text-based files

Importing data using psql

Psql is the pgsql's command-line tool. While one can achieve quite a lot with GUI based database management utilities, psql is very useful when one needs to handle database backups, management and alike via scripting. When there is no GUI installed on the server, psql becomes pretty much the only option so it is worth being familiar with it even if you're not a fan.

In order to import the data in psql we will use a `\COPY` command. This requires us to define the data model for the incoming data first.

Defining the table data model from a text file may be prone to errors that will prevent data from being imported. If for of some reason you are not sure what data types are stored in the particular columns of your source file you can import all the data as text and then re-cast it as required at a later time.

Importing data interactively

In this example we will import the earthquakes data from USGS. So let's fire up psql and connect to the database server:

```
F:\mastering_postgis\chapter02>psql -h localhost -p 5434 -U postgres
```

You should see a similar output:

```
psql (9.5.0)
Type "help" for help.
postgres=#
```

Then we need to connect to the `mastering_postgis` database:

```
postgres=# \c mastering_postgis
```

The following output should be displayed:

```
You are now connected to database "mastering_postgis" as user
"postgres".
mastering_postgis=#
```

In the psql examples I am using `postgres` user. As you may guess, it is a superuser account. This is not the thing you would normally do, but it will keep the examples simple.

In a production environment, using a db user with credentials allowing access to specific resources is a sensible approach.

If you have not had a chance to create our `data_import` schema, let's take care of it now by typing the following command:

```
mastering_postgis=# create schema if not exists data_import;
```

You should see a similar output:

```
NOTICE:  schema "data_import" already exists, skipping
CREATE SCHEMA
```

Once the schema is there, we create the table that will store the data. In order to do so just type or paste the following into psql:

```
create table data_import.earthquakes_csv (
    "time" timestamp with time zone,
    latitude numeric,
    longitude numeric,
    depth numeric,
    mag numeric,
    magType varchar,
    nst numeric,
    gap numeric,
    dmin numeric,
    rms numeric,
    net varchar,
    id varchar,
    updated timestamp with time zone,
    place varchar,
    type varchar,
    horizontalError numeric,
    depthError numeric,
    magError numeric,
    magNst numeric,
    status varchar,
    locationSource varchar,
    magSource varchar
);
```

You should see the following output:

```
mastering_postgis=# create table data_import.earthquakes_csv (
mastering_postgis(# "time" timestamp with time zone,
mastering_postgis(# latitude numeric,
mastering_postgis(# longitude numeric,
mastering_postgis(# depth numeric,
mastering_postgis(# mag numeric,
mastering_postgis(# magType varchar,
mastering_postgis(# nst numeric,
mastering_postgis(# gap numeric,
mastering_postgis(# dmin numeric,
mastering_postgis(# rms numeric,
mastering_postgis(# net varchar,
mastering_postgis(# id varchar,
mastering_postgis(# updated timestamp with time zone,
mastering_postgis(# place varchar,
mastering_postgis(# type varchar,
mastering_postgis(# horizontalError numeric,
mastering_postgis(# depthError numeric,
```

```
mastering_postgis(# magError numeric,
mastering_postgis(# magNst numeric,
mastering_postgis(# status varchar,
mastering_postgis(# locationSource varchar,
mastering_postgis(# magSource varchar
mastering_postgis(# );
CREATE TABLE
```

Now, as we have our data table ready, we can finally get to the import part. The following command should handle importing the data into our newly created table:

```
\copy data_import.earthquakes_csv from data\2.5_day.csv with DELIMITER ','
CSV HEADER
```

You should see a similar output:

```
mastering_postgis=# \copy data_import.earthquakes_csv from data\2.5_day.csv
with DELIMITER ',' CSV HEADER
    COPY 25
```

 If you require a complete reference on the \COPY command, simply type in: \h COPY into the cmd.

While you can customize your data after importing, you may wish to only import a subset of columns in the first place. Unfortunately \COPY command imports all the columns (although you may specify where to put them) so the solution to this would be:

- Create a table that models the input CSV
- Import all the data
- Create a table with a subset of columns
- Copy data over
- Delete the input table

Even though everything said earlier is possible in psql, it requires quite a lot of typing. Because of that we will take care of this scenario in the next stage. Demonstrating the non-interactive psql mode.

Importing data non-interactively

For the non-interactive psql data import example we'll do a bit more than in the interactive mode. We'll:

- Import the full earthquakes dataset
- Select a subset of earthquakes data mentioned in the previous example and insert it into its own table
- Import another dataset - in this case the Ordnance Survey's POIs

Basically the non-interactive usage of psql means we simply provide it with an SQL to execute. This way we can put together many statements without having to execute them one by one.

Once again we will need the data model prior to loading the data, and then a \COPY command will be used.

If you're still in psql, you can execute a script by simply typing:

```
\i path\to\the\script.sql
```

For example:

```
\i F:/mastering_postgis/chapter02/code/data_import_earthquakes.sql
```

You should see a similar output:

```
mastering_postgis-# \i
F:/mastering_postgis/chapter02/code/data_import_earthquakes.sql
    CREATE SCHEMA
    psql:F:/mastering_postgis/chapter02/code/data_import_earthquakes.sql:5:
NOTICE:  table "earthquakes_csv" does not exist, skipping
    DROP TABLE
    CREATE TABLE
    COPY 25
psql:F:/mastering_postgis/chapter02/code/data_import_earthquakes.sql:58:
NOTICE:  table "earthquakes_csv_subset" does not exist, skipping
    DROP TABLE
    SELECT 25
    mastering_postgis-#
```

If you quit psql already, type the following command into cmd:

```
psql -h host -p port -U user -d database -f path\to\the\script.sql
```

For example:

```
psql -h localhost -p 5434 -U postgres -d mastering_postgis -f
F:\mastering_postgis\chapter02\code\data_import_earthquakes.sql
```

You should see a similar output:

```
F:\mastering_postgis\chapter02>psql -h localhost -p 5434 -U postgres -d
mastering_postgis -f
F:\mastering_postgis\chapter02\code\data_import_earthquakes.sql
    psql:F:/mastering_postgis/chapter02/code/data_import_earthquakes.sql:2:
NOTICE:  schema "data_import" already exists, skipping
    CREATE SCHEMA
    DROP TABLE
    CREATE TABLE
    COPY 25
    DROP TABLE
  SELECT 25
```

The script executed earlier is in the book's code repository under Chapter02/code/
data_import_earthquakes.sql.

Loading OS POI data is now a piece of cake. This dataset is in a bit of a different format
though, so it requires slight adjustments. You can review the code in Chapter02/code/
data_import_gb_poi.sql.

Importing data using pgAdmin

In this section we'll import some new data we have not interacted with before - this time
we'll have a look at the Ordnance Survey's address data we obtained in the CSV format.

Depending on the pgAdmin version, the UI may differ a bit. The described
functionality should always be present though.
For the examples involving pgAdmin, screenshots were taken using
pgAdmin III (1.22.2).

PgAdmin's import functionality is basically a wrapper around the \COPY so it does require a data model in order to work. Because of that, let's quickly create a table that will be populated with the imported data. You can do it with the GUI by simply right-clicking a schema node you want to create the table in and choosing **New Object** | **New Table** and then providing all the necessary model definitions in the displayed window:

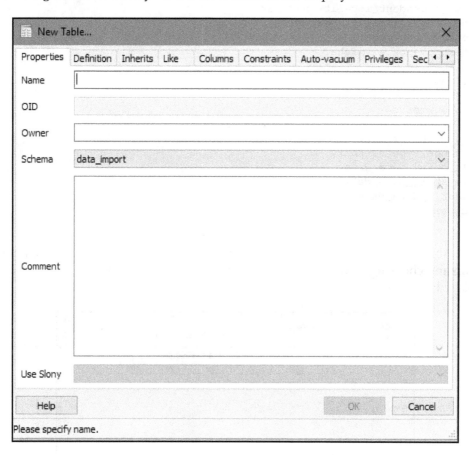

You can also type some SQL which in many cases is a bit quicker:

```
drop table if exists data_import.osgb_addresses;
create table data_import.osgb_addresses(
    uprn bigint,
    os_address_toid varchar,
    udprn integer,
    organisation_name varchar,
    department_name varchar,
    po_box varchar,
```

```
    sub_building_name varchar,
    building_name varchar,
    building_number varchar,
    dependent_thoroughfare varchar,
    thoroughfare varchar,
    post_town varchar,
    dbl_dependent_locality varchar,
    dependent_locality varchar,
    postcode varchar,
    postcode_type varchar,
    x numeric,
    y numeric,
    lat numeric,
    lon numeric,
    rpc numeric,
    country varchar,
    change_type varchar,
    la_start_date date,
    rm_start_date date,
    last_update_date date,
    class varchar
);
```

Once our table is ready, importing data is just a matter of right clicking the table node in **PgAdmin** and choosing **Import**. An import wizard that assists with the import process will be displayed:

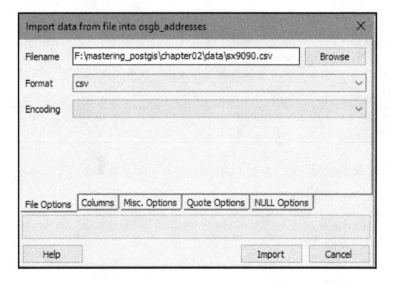

All the earlier could obviously be achieved with pure SQL and in fact we have done this already in the previous section on importing data in psql in non-interactive mode. You can review the SQL code available in Chapter02/code for details.

Extracting spatial information from flat data

As we have some flat data already in our database, it's time to extract the spatial information. So far all the datasets, used Cartesian coordinate systems so our job is really straightforward:

```
drop table if exists data_import.earthquakes_subset_with_geom;
select
    id,
    "time",
    depth,
    mag,
    magtype,
    place,Points of Interest in TXT format
    ST_SetSRID(ST_Point(longitude, latitude), 4326) as geom
into data_import.earthquakes_subset_with_geom
from data_import.earthquakes_csv;
```

This example extracts a subset of data and puts data into a new table with coordinates being expressed as a geometry type, rather than two columns with numeric data appropriate for Lon and Lat.

In order to quickly preview the data, we dump the table's content to KML using ogr2ogr (this is a little spoiler on the next chapter on exporting the data from PostGIS indeed):

```
ogr2ogr -f "KML" earthquakes_from_postgis.kml PG:"host=localhost port=5434
user=postgres dbname=mastering_postgis"
data_import.earthquakes_subset_with_geom -t_srs EPSG:4326
```

Such KML can be viewed for example in Google Earth (you can use the original KML downloaded from USGS just as a cross check for the output data):

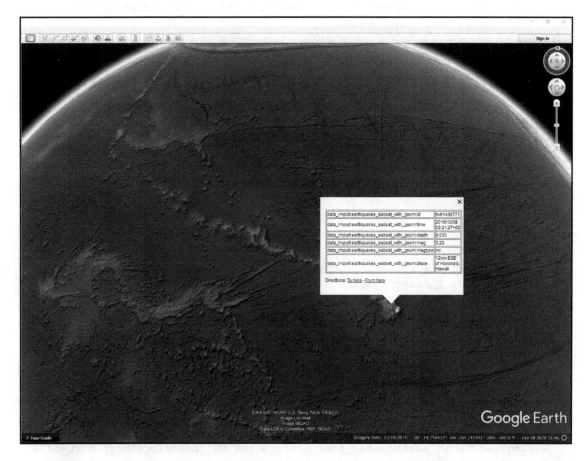

 More examples of extracting the spatial data from different formats are addressed in the ETL chapter.

Importing shape files using shp2pgsql

ESRI **shapefile** (**SHP**) is still the most common exchange format for sharing GIS data. The format itself is made of a few files such as SHP, SHX, DBF, andPRJ, where the first three are the required files and the file with projection information is not obligatory.

The standard PostGIS tool for loading shapefiles is `shp2pgsql` - you will find it in the bin folder of your postgres installation. `shp2pgsql` is a command-line utility that can either extract the shapefile data into SQL or pipe the output directly into psql (we'll see both approaches). `shp2pgsql` also has a GUI version that can be accessed directly in PgAdmin.

In this example, we'll use some `NaturalEarth` shapefiles we downloaded earlier. We will import the coastlines shapefile using the CMD version of `shp2pgsql` and then we'll add land masses using the GUI version.

shp2pgsql in cmd

`shp2pgsql` has an extensive list of parameters that can be accessed by simply typing `shp2pgsql` in the cmd. We will not use all the options but rather explain the most common ones.

The basic usage of the utility is as follows:

```
shp2pgsql [<options>] <shapefile> [[<schema>.]<table>]
```

For example:

```
shp2pgsql -s 4326 ne_110m_coastline data_import.ne_coastline
```

Basically you specify what shapefile you want to import and where. If you executed the earlier command, you would just see the `shp2pgsql` plain SQL output logged to the console, similar to this:

```
...
INSERT INTO "data_import"."ne_coastline" ("scalerank","featurecla",geom)
VALUES
('3','Country','0105000020E6100000010000000102000000060000006666666666A65AC
0676666666666652407130AD7A3505AC0295C8FC2F5685240000000000205AC07B14AE47E1
5A5240B91E85EB51585AC07130AD7A33052405C8FC2F528BC5AC03E0AD7A3705D524066666
66666A65AC067666666666652400');
    COMMIT;
    ANALYZE "data_import"."ne_coastline";
```

So basically, we need to do something with the utility output in order to make use of it. Let's save the output to an SQL file and let psql read it first:

```
shp2pgsql -s 4326 ne_110m_coastline data_import.ne_coastline >
ne_110m_coastline.sql
psql -h localhost -p 5434 -U postgres -d mastering_postgis -f
ne_110m_coastline.sql
```

You should see a similar output:

```
SET
SET
BEGIN
CREATE TABLE
ALTER TABLE


                         addgeometrycolumn
---------------------------------------------------------------------
data_import.ne_coastline.geom SRID:4326 TYPE:MULTILINESTRING DIMS:2
(1 row)

INSERT 0 1
...
INSERT 0 1
COMMIT
ANALYZE
```

I suggest you have a look at the generated SQL so you get an idea of what is actually happening behind the scenes.

Now let's pipe the shp2pgsql output directly to psql:

```
shp2pgsql -s 4326 ne_110m_coastline data_import.ne_coastline | psql -h
localhost -p 5434 -U postgres -d mastering_postgis
```

The cmd output will be exactly the same as the one we have already seen when reading data from the SQL file.

 You will have to drop the data_import.ne_coastline table before importing the data again; otherwise the command in its current shape will fail.

There are some shp2pgsql options that are worth remembering:

- -s SRID: Specifies the shp data projection identifier. When used in the following form: -s SRID1:SRID2 makes the shp2pgsql apply a coordinate system transformation, so projection of the data changes is required.
- -p: Turns on the 'prepare' mode - only a table definition is output.
- -d: Drops and recreates a table.
- -a: Appends data to the existing table, provided its schema is exactly the same as the schema of the incoming data.

- −g: Allows specifying of the geometry column name; the default is geom (or geog if you decide to use geography with the -G param).
- −m <filename>: Specifies a file name that contains column mapping for the DBF file. This way, you can remap dbf column names to your preference.
- −n: Only imports DBF and no spatial data.

Importing data with SRID transformation: −s SRID1:SRID2.

The shp2pgsql GUI version

shp2pgsql also has a GUI version. In order to use it, when in PgAdmin, simply choose Plugins | PostGIS Shapefile and DBF loader 2.2 (the version may vary); the following import wizard will be displayed:

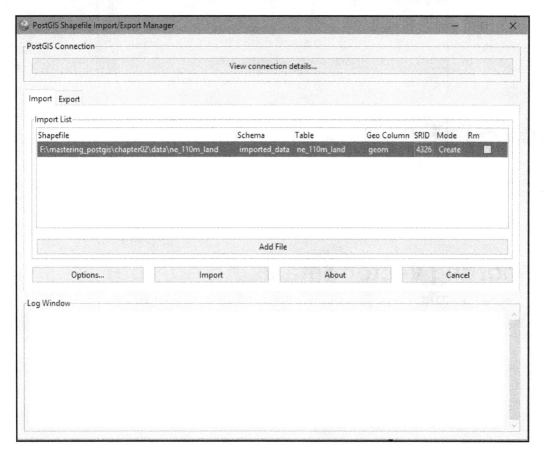

In pgAdmin 4, accessing the shapefile loader GUI may not be so obvious. To trigger the tool, try typing `shp2pgsql-gui` in the shell/command line.

Similar to the cmd version of the utility, you can specify the schema, and you should specify the SRID.

The nice thing about the GUI version of `shp2pgsql` is that it lets one import multiple files at once.

In the options dialogue, you can specify data encoding, decide whether or not you would like to create a spatial index after importing, choose geography over geometry, and so on:

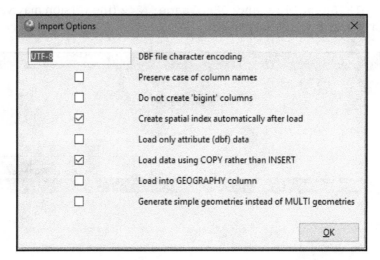

Importing vector data using ogr2ogr

`ogr2ogr` is the GDAL's vector transform utility. It is - not without reason - considered a Swiss Army knife for vector transformations. Despite its size, `ogr2ogr` can handle a wide range of formats and this makes it a really worthy tool.

We'll use `ogr2ogr` to import a few data formats other than SHP, although `ogr2ogr` can obviously import SHP too. For this scenario, we'll use some data downloaded earlier:

- OS GB address base in GML format
- OS GB code point polygons in MapInof MIF & TAB formats
- USGS earthquakes in KML format

Some of the most common ogr2ogr params are:

- `-f`: The format of the output (when importing to PostGIS it will be PostgreSQL).
- `-nln`: Assigns a name to the layer. In the case of importing the data to PostGIS this will be the table name.
- `-select`: Lets you specify a comma separated list of columns to pick.
- `-where`: Lets you specify a sql like query to filter out the data.
- `append`: Appends data to the output dataset.
- `overwrite`: Overwrites the output datasource - in case of PostgreSQL it will drop and re-create a table.
- `s_srs`: Specifies the input SRID.
- `t_srs`: Transforms coordinates to the specified SRID.
- `a_srs`: Specifies the output SRID.
  - `lco NAME=VALUE`: Layer creation options - these are driver specific; for pgsql options, see `http://www.gdal.org/drv_pg.html`. The most commonly used layer creation options are:
  - `LAUNDER`: This defaults to `YES`. It is responsible for converting column names into pgsql compatible ones (lower case, underscores).
  - `PRECISION`: This defaults to `YES`. It is responsible for using numeric and char types over float and varchar.
  - `GEOMETRY_NAME`: Defaults to `wkb_geometry`.

For a full list of `ogr2ogr` params, just type `ogr2ogr`.
`Ogr2ogr` has an accompanying utility called **ogrinfo**. This tool lets one inspect the metadata of a dataset. Verifying the metadata of any dataset prior to working with it is considered good practice and one should get into the habit of always using it before importing or exporting the data.

Importing GML

Let's start with importing the GML of the OS GB address base. First we'll see what data we're dealing with exactly:

```
ogrinfo sx9090.gml
```

The following should be the output:

```
Had to open data source read-only.
INFO: Open of `sx9090.gml'
      using driver `GML' successful.
1: Address (Point)
```

We can then review the layer information:

```
ogrinfo sx9090.gml Address -so
```

You should see a similar output:

```
Had to open data source read-only.
INFO: Open of `sx9090.gml'
      using driver `GML' successful.

Layer name: Address
Geometry: Point
Feature Count: 42861
Extent: (-3.560100, 50.699470) - (-3.488340, 50.744770)
Layer SRS WKT:
GEOGCS["ETRS89",
    DATUM["European_Terrestrial_Reference_System_1989",
        SPHEROID["GRS 1980",6378137,298.257222101,
            AUTHORITY["EPSG","7019"]],
        TOWGS84[0,0,0,0,0,0,0],
        AUTHORITY["EPSG","6258"]],
    PRIMEM["Greenwich",0,
        AUTHORITY["EPSG","8901"]],
    UNIT["degree",0.0174532925199433,
        AUTHORITY["EPSG","9122"]],
    AUTHORITY["EPSG","4258"]]
gml_id: String (0.0)
uprn: Real (0.0)
osAddressTOID: String (20.0)
udprn: Integer (0.0)
subBuildingName: String (25.0)
buildingName: String (36.0)
thoroughfare: String (27.0)
postTown: String (6.0)
```

```
postcode: String (7.0)
postcodeType: String (1.0)
rpc: Integer (0.0)
country: String (1.0)
changeType: String (1.0)
laStartDate: String (10.0)
rmStartDate: String (10.0)
lastUpdateDate: String (10.0)
class: String (1.0)
buildingNumber: Integer (0.0)
dependentLocality: String (27.0)
organisationName: String (55.0)
dependentThoroughfare: String (27.0)
poBoxNumber: Integer (0.0)
doubleDependentLocality: String (21.0)
departmentName: String (37.0)
```

 -so param makes ogrinfo display the data summary only; otherwise, info on a full dataset would be displayed.

Once we're ready to import the data, let's execute the following command:

```
ogr2ogr -f "PostgreSQL" PG:"host=localhost port=5434 user=postgres
dbname=mastering_postgis" sx9090.gml -nln data_import.osgb_address_base_gml
-geomfield geom
```

At this stage, the address GML should be available in our database.

 We did not specify the SRID of the GML data. This is because this information is present in GML and the utility picks it up automatically.

Importing MIF and TAB

Both MIF and TAB are MapInfo formats. TAB is the default format that contains formatting, while MIF is the interchange format.

We'll start with reviewing metadata:

```
ogrinfo EX_sample.mif
```

And then:

```
ogrinfo EX_sample.mif EX_Sample -so
Had to open data source read-only.
INFO: Open of `EX_sample.mif'
      using driver `MapInfo File' successful.

Layer name: EX_sample
Geometry: Unknown (any)
Feature Count: 4142
Extent: (281282.800000, 85614.570000) - (300012.000000,
100272.000000)
Layer SRS WKT:
PROJCS["unnamed",
    GEOGCS["unnamed",
        DATUM["OSGB_1936",
            SPHEROID["Airy 1930",6377563.396,299.3249646],
            TOWGS84[375,-111,431,-0,-0,-0,0]],
        PRIMEM["Greenwich",0],
        UNIT["degree",0.0174532925199433]],
    PROJECTION["Transverse_Mercator"],
    PARAMETER["latitude_of_origin",49],
    PARAMETER["central_meridian",-2],
    PARAMETER["scale_factor",0.9996012717],
    PARAMETER["false_easting",400000],
    PARAMETER["false_northing",-100000],
    UNIT["Meter",1]]
POSTCODE: String (8.0)
UPP: String (20.0)
PC_AREA: String (2.0)
```

Please note that **ogrinfo** projection metadata for our MIF file does not specify the EPSG code. This is fine, as the projection definition is present. But it will result in ogr2ogr creating a new entry in the `spatial_ref_sys`, which is not too good, as we'll end up with the wrong `coordsys` identifiers; the coordinate reference id will be the next available.

This is because `ogr2ogr` expands the coordinate reference into a WKT string and then does a string comparison against the `coordsys` identifiers definitions in the `spatial_ref_sys table`; minor differences in formatting or precision will result in `ogr2ogr` failing to match `coordsys`. In such a scenario, a new entry will be created; for example, if you happen to use the *EPSG:3857* coordinate system and the system's definition is slightly different and cannot be matched, the assigned SRID will not be 3857, but the next available ID will be chosen.

A solution to this is to specify the exact coordinate system; `ogr2ogr` should output the data via the `a_srs` parameter.

Once ready, we can import the data:

```
ogr2ogr -f "PostgreSQL" PG:"host=localhost port=5434 user=postgres
dbname=mastering_postgis" EX_sample.mif -nln
data_import.osgb_code_point_polygons_mif -lco GEOMETRY_NAME=geom -a_srs
EPSG:27700
```

If you followed the very same procedure for TAB file and loaded the data, both datasets are now in their own tables in the `data_import` schema:

```
ogr2ogr -f "PostgreSQL" PG:"host=localhost port=5434 user=postgres
dbname=mastering_postgis" EX_sample.tab -nln
data_import.osgb_code_point_polygons_tab -lco GEOMETRY_NAME=geom -a_srs
EPSG:27700
```

Importing KML

As usual, we'll start with the dataset's metadata checkup:

```
ogrinfo 2.5_day_age.kml
```

The output shows that there is more than one layer:

```
INFO: Open of `2.5_day_age.kml'
      using driver `LIBKML' successful.
1: Magnitude 5
2: Magnitude 4
3: Magnitude 3
4: Magnitude 2
```

Therefore, in order to review metadata for each layer at once, the following command should be used:

```
ogrinfo 2.5_day_age.kml -al -so
```

The output of the previous command is rather longish, so we'll truncate it a bit and only show the info for the first layer:

```
INFO: Open of `2.5_day_age.kml'
      using driver `LIBKML' successful.

Layer name: Magnitude 5
Geometry: Unknown (any)
Feature Count: 2
Extent: (-101.000100, -36.056300) - (120.706400, 13.588200)
Layer SRS WKT:
GEOGCS["WGS 84",
    DATUM["WGS_1984",
        SPHEROID["WGS 84",6378137,298.257223563,
            AUTHORITY["EPSG","7030"]],
        TOWGS84[0,0,0,0,0,0,0],
        AUTHORITY["EPSG","6326"]],
    PRIMEM["Greenwich",0,
        AUTHORITY["EPSG","8901"]],
    UNIT["degree",0.0174532925199433,
        AUTHORITY["EPSG","9108"]],
    AUTHORITY["EPSG","4326"]]
Name: String (0.0)
description: String (0.0)
timestamp: DateTime (0.0)
begin: DateTime (0.0)
end: DateTime (0.0)
altitudeMode: String (0.0)
tessellate: Integer (0.0)
extrude: Integer (0.0)
visibility: Integer (0.0)
drawOrder: Integer (0.0)
icon: String (0.0)
snippet: String (0.0)
```

This time, EPSG information is available, so we do not have to worry; ogr2ogr will create extra SRID definition in the database.

Once we've confirmed that this is the exact dataset we'd like to import, we can continue with the following command:

```
ogr2ogr -f "PostgreSQL" PG:"host=localhost port=5434 user=postgres
dbname=mastering_postgis" 2.5_day_age.kml -nln
data_import.usgs_earthquakes_kml -lco GEOMETRY_NAME=geom -append
```

> Note the append param in the command earlier. This is required because our KML has more than one layer and ogr2ogr is importing them one by one. Without the append param, only the first layer would be imported and then ogr2ogr would fail with a similar output:
> FAILED: Layer data_import.usgs_earthquakes_kml already exists, and -append not specified.
> Consider using -append, or -overwrite.
> ERROR 1: Terminating translation prematurely after failed translation of layer Magnitude 4 (use -skipfailures to skip errors)

The cmd output should be similar to:

```
WARNING: Layer creation options ignored since an existing layer is
      being appended to.
WARNING: Layer creation options ignored since an existing layer is
      being appended to.
WARNING: Layer creation options ignored since an existing layer is
      being appended to.
```

At this stage, the KML dataset should have made it to our PostGIS database.

ogr2ogr GUI (Windows only)

For those preferring GUI over CMD, there is an alternative to plain old ogr2ogr--ogr2gui available from http://www.ogr2gui.ca/.

Simply download the required archive, extract it, and launch the appropriate .exe. After having played with ogr2ogr a bit already, the GUI should be rather self-explanatory:

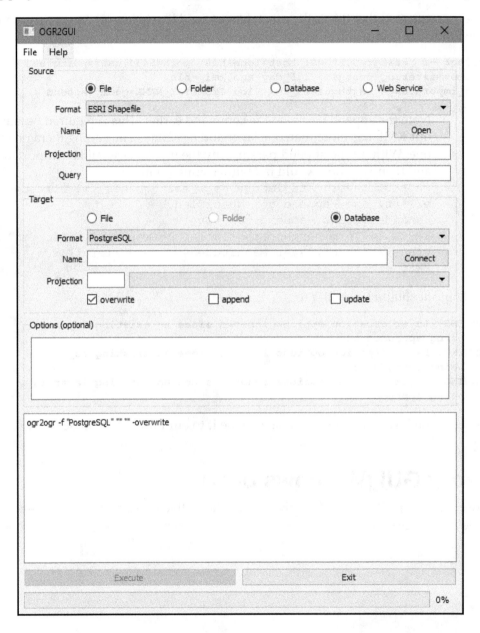

Importing data using GIS clients

Many GIS software packages can directly connect to databases for reading and writing the data; from our perspective, they are just more database clients. In this section, we'll have a quick look at the very well-known QGIS and the certainly less popular but very powerful Manifold GIS.

Both can export data to databases and read it back. QGIS has a specialized PostGIS export module called **SPIT**; Manifold's export facility is built in into the GUI and follows the same export routines as other formats handled by the software.

Exporting a shapefile to PostGIS using QGIS and SPIT

QGIS offers a PostGIS export module called SPIT. You can access it by choosing
`Database\Spit\Import Shapefiles to PostGIS:`

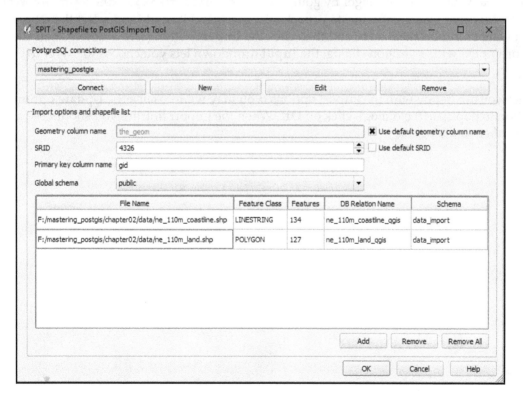

SPIT's GUI is very clear and easy to understand. You can import many files at once and you can specify the destination schema and table names. If required you can change the default Spit's geometry name (the_geom) to your liking. SRID also can be changed, but it will be applied to all the imported files. Once you provide all the required information and click on **OK**, a progress window is displayed and the data is exported to PostGIS.

 In newer versions of QGIS, you may not find SPIT anymore. In such cases, you can use a DbManager instead.

Exporting shapefile to PostGIS using QGIS and DbManager

Before using DbManager, you should load a shapefile you want to import to a database.

When ready, launch DbManager by going to Database\DB Manager\DB Manager. When the DbManager UI displays, locate your database, expand its node and select the schema you want to import the data to and then click the **Import** button (arrow down). You should be prompted with an import dialog; the **Input** dropdown lets you choose the layer to import.

Once you are happy with the import options (you may want to click the **Update options** button to populate the dialog), click on **OK**. When the import finishes you should see a confirmation dialogbox:

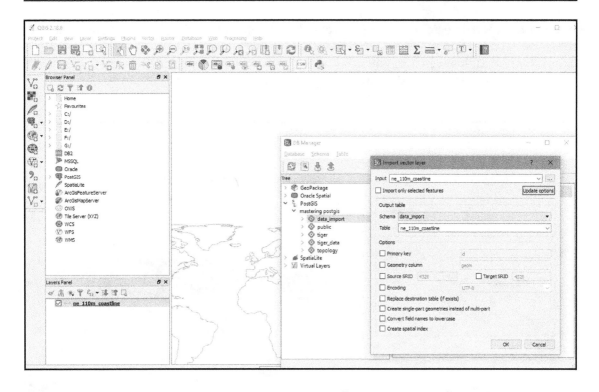

Exporting spatial data to PostGIS from Manifold GIS

In order to export the data, Manifold needs to import it internally first. This means that we do not export a particular format but rather we simply export spatial data. Once you bring a shapefile into Manifold, MapInfo TAB or MIF, SQL Server spatial table or any other supported vector format, exporting it to PostGIS is exactly the same for all of them.

In order to export a vector component, right-click on its node in the project tree and choose the export option. You will see an export dialog, where you should pick the appropriate export format; in this scenario, you need to choose **Data Source**. You will then be presented with a data source setup window, where you can either pick an existing connection or configure a new one:

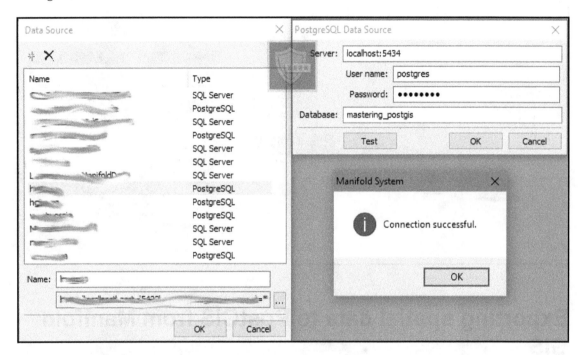

Once you choose the appropriate connection, you can then set up the actual export parameters:

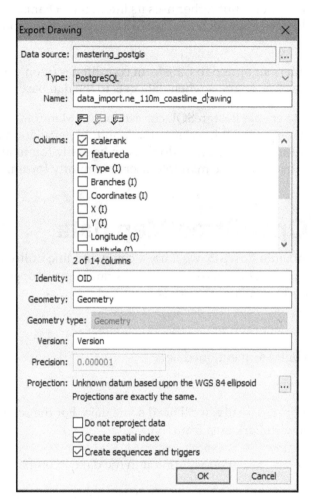

You can choose which columns you want to export, and the name of the identity, geometry, and version columns.

There is a minor inconvenience with the exporter: it does not allow for adjusting of the destination schema and always exports to the public schema.

Manifold tries to find the PostGIS projection that best matches the Manifold projection. Unfortunately, it is not always possible as Manifold as such does not rely on EPSG coordinates systems definitions, but rather uses its internal mechanisms for handling projections. If Manifold does not match the PostGIS side projection, you can select it manually.

Export dialogue also offers an option to transform coordinates upon export and create indexes and CREATE/UPDATE triggers when data gets to the database.

 In order to enable PostgreSQL connections in Manifold, you may have to copy some PgSQL DLLs over to the Manifold installation directory. The exact information on how to do this can be easily found at georeference.org, the manifold user community forum.

Importing OpenStreetMap data

For importing OSM data into PostGIS, we'll use a command line utility called osm2pgsql. Apparently, making a Linux build of osm2pgsql is straightforward; getting one that runs on Windows may require some more effort as described here:
https://github.com/openstreetmap/osm2pgsql/issues/17,
https://github.com/openstreetmap/osm2pgsql/issues/472.

I have used a Cygwin build as mentioned here:

http://wiki.openstreetmap.org/wiki/Osm2pgsql#Cygwin

Once we have the osm2pgsql ready, we'll need some data. For the sake of simplicity, I have downloaded the Greenwich Park area from
https://www.openstreetmap.org/export#map=16/51.4766/0.0003 and saved the file as greenwich_observatory.osm (you will find it in the data accompanying this chapter).

The downloaded file is actually an XML file. Do have a look what's inside to get an idea of the data osm2pgsql is dealing with.

In order to take advantage of the OSM tags used to describe the data, we will need the PostgreSQL hstore extension. Basically it allows for storing key-value pairs in a column, so data with flexible schema can easily be stored. In order to install it, you need to execute the following query in either PgAdmin or psql:

```
CREATE EXTENSION hstore;
```

In order to import OSM data, issue the following command, making sure you adjust the paths and db connection details to your environment:

```
osm2pgsql.exe -H localhost -P 5434 -U postgres -W -d mastering_postgis -S
default.style ../data/greenwich_observatory.osm -hstore
```

 If you happen to receive a message such as **Default style not found**, please make sure to provide a valid path to the styles definition such as /usr/share/osm2pgsql/default.style.

You should see a similar output:

```
osm2pgsql SVN version 0.85.0 (64bit id space)
Password:
Using projection SRS 900913 (Spherical Mercator)
Setting up table: planet_osm_point
NOTICE:  table "planet_osm_point" does not exist, skipping
NOTICE:  table "planet_osm_point_tmp" does not exist, skipping
Setting up table: planet_osm_line
NOTICE:  table "planet_osm_line" does not exist, skipping
NOTICE:  table "planet_osm_line_tmp" does not exist, skipping
Setting up table: planet_osm_polygon
NOTICE:  table "planet_osm_polygon" does not exist, skipping
NOTICE:  table "planet_osm_polygon_tmp" does not exist, skipping
Setting up table: planet_osm_roads
NOTICE:  table "planet_osm_roads" does not exist, skipping
NOTICE:  table "planet_osm_roads_tmp" does not exist, skipping
Using built-in tag processing pipeline
Allocating memory for sparse node cache
Node-cache: cache=800MB, maxblocks=0*102400, allocation method=8192
Mid: Ram, scale=100
!! You are running this on 32bit system, so at most
!! 3GB of RAM can be used. If you encounter unexpected
!! exceptions during import, you should try running in slim
!! mode using parameter -s.

Reading in file: ../data/greenwich_observatory.osm
Processing: Node(4k 4.7k/s) Way(0k 0.55k/s) Relation(41 41.00/s)   parse
time: 0s
Node stats: total(4654), max(4268388189) in 0s
Way stats: total(546), max(420504897) in 0s
Relation stats: total(41), max(6096780) in 0s
Committing transaction for planet_osm_point
Committing transaction for planet_osm_line
Committing transaction for planet_osm_polygon
Committing transaction for planet_osm_roads
Writing relation (41)
```

```
Sorting data and creating indexes for planet_osm_point
Analyzing planet_osm_point finished
Sorting data and creating indexes for planet_osm_line
Sorting data and creating indexes for planet_osm_polygon
Analyzing planet_osm_line finished
node cache: stored: 4654(100.00%), storage efficiency: 50.00% (dense
blocks: 0, sparse nodes: 4654), hit rate: 2.00%
Sorting data and creating indexes for planet_osm_roads
Analyzing planet_osm_polygon finished
Analyzing planet_osm_roads finished
Copying planet_osm_point to cluster by geometry finished
Creating geometry index on  planet_osm_point
Creating indexes on  planet_osm_point finished
All indexes on  planet_osm_point created  in 0s
Completed planet_osm_point
Copying planet_osm_line to cluster by geometry finished
Creating geometry index on  planet_osm_line
Creating indexes on  planet_osm_line finished
Copying planet_osm_polygon to cluster by geometry finished
Creating geometry index on  planet_osm_polygon
All indexes on  planet_osm_line created  in 0s
Completed planet_osm_line
Creating indexes on  planet_osm_polygon finished
Copying planet_osm_roads to cluster by geometry finished
Creating geometry index on  planet_osm_roads
All indexes on  planet_osm_polygon created  in 0s
Completed planet_osm_polygon
Creating indexes on  planet_osm_roads finished
All indexes on  planet_osm_roads created  in 0s
Completed planet_osm_roads
Osm2pgsql took 1s overall
```

At this stage, you should have the OSM data imported to the public schema. Thanks to using the `hstore` datatype for tags column, we can now do the following type of queries:

```
select name FROM planet_osm_point where ((tags->'memorial') = 'stone');
```

When executed in psql with the dataset used in this example, you should see the following output:

```
            name
-------------------------------
 Prime Meridian of the World
(1 row)
```

You may want to index the tags columns in order to optimize the query performance.

Connecting to external data sources with foreign data wrappers

Since PostgreSQL 9.1, one can use **Foreign Data Wrappers** (**FDW**) in order to connect to the external data sources that are then treated as they were local tables. More information can be found on the PostgreSQL wiki:

https://wiki.postgresql.org/wiki/Foreign_data_wrappers.

Querying the external files or databases seems to be standard these days. But how about querying WFS services or OSM directly? Now, this sounds cool, doesn't it? You should certainly have a look at some of the clever GEO data wrappers:

- ogr_fdw: https://github.com/pramsey/pgsql-ogr-fdw
- osm_pbf_fdw: https://github.com/vpikulik/postgres_osm_pbf_fdw

In this example, we'll use ogr_fdw to connect to some external data sources. Starting with PostGIS 2.2, it is a part of the bundle and there is no need to install it as it should already be available.

 Examples shown in this section can be executed in both psql or in PgAdmin.

Connecting to SQL Server Spatial

First we need to create a server:

```
CREATE SERVER fdw_sqlserver_test
  FOREIGN DATA WRAPPER ogr_fdw
  OPTIONS (
    datasource
'MSSQL:server=CM_DOM\MSSQLSERVER12;database=hgis;UID=postgres_fdw;PWD=postg
res_fdw',
    format 'MSSQLSpatial');
```

You may have noticed I have created a `postgres_fdw` user with the same password.

If you're using Postgre SQL 9.5+, you can use the `IMPORT SCHEMA` command:

```
IMPORT FOREIGN SCHEMA "dbo.Wig100_skorowidz"
FROM SERVER fdw_sqlserver_test INTO data_linked;
```

Otherwise you will have to specify the table schema explicitly:

```
CREATE FOREIGN TABLE data_linked.dbo_wig100_skorowidz
    (fid integer ,
     geom public.geometry ,
     oid integer ,
     gid integer ,
     version integer ,
     godlo character varying ,
     nazwa character varying ,
     nazwa2 character varying ,
     kalibracja character varying ,
     pas real ,
     slup real )
SERVER fdw_sqlserver_test
OPTIONS (layer 'dbo.Wig100_skorowidz');
```

 By default, PgAdmin does not display foreign tables, so you may have to go to **File** | **Options** and tick the **Foreign Tables** checkbox in the Browser node. In PgAdmin 4, foreign tables seem to be visible by default.

At this stage, you should be able to query the foreign table as if it was local.

Connecting to WFS service

This example is based on the `ogr_fwd` documentation, so it only shows the required stuff. A full example can be reviewed here:

```
https://github.com/robe2/pgsql-ogr-fdw
```

First let's create a foreign server:

```
CREATE SERVER fdw_wfs_test_opengeo
  FOREIGN DATA WRAPPER ogr_fdw
  OPTIONS (
    datasource 'WFS:http://demo.opengeo.org/geoserver/wfs',
    format 'WFS');
```

Automagically bring in the schema:

```
IMPORT FOREIGN SCHEMA "topp:tasmania_cities"
FROM SERVER fdw_wfs_test_opengeo INTO data_linked;
```

And issue a query against the foreign WFS table:

```
select city_name from data_linked.topp_tasmania_cities;
```

Since this dataset contains only one record, our result should be Hobart.

Loading rasters using raster2pgsql

`raster2pgsql` is the default tool for importing rasters to PostGIS. Even though GDAL itself does not provide means to load rasters to the database, `raster2pgsql` is compiled as a part of PostGIS and therefore supports the very same formats as the GDAL version appropriate for given version of PostGIS.

`raster2pgsql` is a command-line tool. In order to review its parameters, simply type in the console:

```
raster2pgsql
```

While taking a while to get familiar with the `raster2pgsql` help is an advised approach, here are some params that worth highlighting:

- `-G`: Prints a list of GDAL formats supported by the given version of the utility; the list is likely to be extensive.
- `-s`: Sets the SRID of the imported raster.
- `-t`: Tile size - expressed as width x height. If not provided, a default is worked out automatically in the range of 32-100 so it best matches the raster dimensions. It is worth remembering that when importing multiple files, tiles will be computed for the first raster and then applied to others.

- -P: Pads tiles right / bottom, so all the tiles have the same dimensions.
- -d|a|c|p: These options are mutually exclusive:
    - d: Drops and creates a table.
    - a: Appends data to an existing table.
    - c: Creates a new table.
    - p: Turns on *prepare* mode. So no importing is done; only a table is created.
- -F: A column with raster name will be added.
- -l: Comma-separated overviews; creates overview tables named o_<overview_factor>_raster_table_name.
- -I: Creates GIST spatial index on the raster column.
- -C: Sets the standard constraints on the raster column after the raster is imported.

For the examples used in this section, we'll use Natural Earth's 50M Gray Earth raster.

As you remember, ogr2ogr has a ogrinfo tool that can be used to obtain the information on a vector dataset. GDAL's equivalent for raster files is called gdalinfo and is as worthy as its vector brother:

```
gdalinfo GRAY_50M_SR_OB.tif
```

You should get a similar output:

```
Driver: GTiff/GeoTIFF
Files: GRAY_50M_SR_OB.tif
       GRAY_50M_SR_OB.tfw
Size is 10800, 5400
Coordinate System is:
GEOGCS["WGS 84",
    DATUM["WGS_1984",
        SPHEROID["WGS 84",6378137,298.257223563,
            AUTHORITY["EPSG","7030"]],
        AUTHORITY["EPSG","6326"]],
    PRIMEM["Greenwich",0],
    UNIT["degree",0.0174532925199433],
    AUTHORITY["EPSG","4326"]]
Origin = (-179.999999999999970,90.000000000000000)
Pixel Size = (0.033333333333330,-0.033333333333330)
Metadata:
  AREA_OR_POINT=Area
  TIFFTAG_DATETIME=2014:10:18 09:28:20
  TIFFTAG_RESOLUTIONUNIT=2 (pixels/inch)
  TIFFTAG_SOFTWARE=Adobe Photoshop CC 2014 (Macintosh)
```

```
  TIFFTAG_XRESOLUTION=342.85699
  TIFFTAG_YRESOLUTION=342.85699
Image Structure Metadata:
  INTERLEAVE=BAND
Corner Coordinates:
Upper Left  (-180.0000000,  90.0000000) (180d 0' 0.00"W, 90d 0' 0.00"N)
Lower Left  (-180.0000000, -90.0000000) (180d 0' 0.00"W, 90d 0' 0.00"S)
Upper Right ( 180.0000000,  90.0000000) (180d 0' 0.00"E, 90d 0' 0.00"N)
Lower Right ( 180.0000000, -90.0000000) (180d 0' 0.00"E, 90d 0' 0.00"S)
Center      (  -0.0000000,   0.0000000) (  0d 0' 0.00"W,  0d 0' 0.00"N)
Band 1 Block=10800x1 Type=Byte, ColorInterp=Gray
```

Before we get down to importing the raster, let's splits into four parts using `gdalwarp` utility. This way, we'll be able to show how to import a single raster and a set of rasters:

```
gdalwarp -s_srs EPSG:4326 -t_srs EPSG:4326 -te -180 -90 0 0
GRAY_50M_SR_OB.tif gray_50m_partial_bl.tif
gdalwarp -s_srs EPSG:4326 -t_srs EPSG:4326 -te -180 0 0 90
GRAY_50M_SR_OB.tif gray_50m_partial_tl.tif
gdalwarp -s_srs EPSG:4326 -t_srs EPSG:4326 -te 0 -90 180 0
GRAY_50M_SR_OB.tif gray_50m_partial_br.tif
gdalwarp -s_srs EPSG:4326 -t_srs EPSG:4326 -te 0 0 180 90
GRAY_50M_SR_OB.tif gray_50m_partial_tr.tif
```

For each command, you should see a similar output:

```
Creating output file that is 5400P x 2700L.
Processing input file GRAY_50M_SR_OB.tif.
0...10...20...30...40...50...60...70...80...90...100 - done.
```

Having prepared the data, we can now move onto importing it.

Importing a single raster

In order to import a single raster file, let's issue the following command:

```
raster2pgsql -s 4326 -C -l 2,4 -I -F -t 2700x2700 gray_50m_sr_ob.tif
data_import.gray_50m_sr_ob | psql -h localhost -p 5434 -U postgres -d
mastering_postgis
```

You should see a similar output:

```
Processing 1/1: gray_50m_sr_ob.tif
BEGIN
CREATE TABLE
CREATE TABLE
CREATE TABLE
INSERT 0 1
(...)
INSERT 0 1
CREATE INDEX
ANALYZE
CREATE INDEX
ANALYZE
CREATE INDEX
ANALYZE
NOTICE:  Adding SRID constraint
CONTEXT:  PL/pgSQL function
addrasterconstraints(name,name,name,boolean,boolean,boolean,boolean,boolean
,boolean,boolean,boolean,boolean,boolean,boolean,boolean) line 53 at RETURN
NOTICE:  Adding scale-X constraint
(...)
---------------------
 t
(1 row)

 addoverviewconstraints
-------------------------
 t
(1 row)

 addoverviewconstraints
-------------------------
 t
(1 row)

COMMIT
```

The executed command created 3 tables: the main raster table
called data_import.gray_50m_sr_ob and two overview tables
called data_import.o_2_gray_50m_sr_ob and data_import.o_4_gray_50m_sr_ob.
The command also created the GIST index and brought in the filename. The raster has been
split into tiles of 2700 x 2700 pixels.

Importing multiple rasters

Let's import a directory of rasters now. We have four files with the file name mask `gray_50m_partial*.tif`. In order to import all the files at once, we'll issue the following command:

```
raster2pgsql -s 4326 -C -l 2,4 -I -F -t 2700x2700 gray_50m_partial*.tif
data_import.gray_50m_partial | psql -h localhost -p 5434 -U postgres -d
mastering_postgis
```

You should see a similar output:

```
Processing 1/4: gray_50m_partial_bl.tif
BEGIN
CREATE TABLE
CREATE TABLE
CREATE TABLE
INSERT 0 1
INSERT 0 1
INSERT 0 1
Processing 2/4: gray_50m_partial_br.tif
(...)
Processing 3/4: gray_50m_partial_tl.tif
(...)
Processing 4/4: gray_50m_partial_tr.tif
(...)
CONTEXT:  PL/pgSQL function
addrasterconstraints(name,name,name,boolean,boolean,boolean,boolean,boolean
,boolean,boolean,boolean,boolean,boolean,boolean,boolean) line 53 at RETURN
NOTICE:  Adding maximum extent constraint
CONTEXT:  PL/pgSQL function
addrasterconstraints(name,name,name,boolean,boolean,boolean,boolean,boolean
,boolean,boolean,boolean,boolean,boolean,boolean,boolean) line 53 at RETURN
 addrasterconstraints
-----------------------
 t
(1 row)


 addoverviewconstraints
-----------------------
 t
(1 row)


 addoverviewconstraints
-----------------------
 t
(1 row)
```

```
COMMIT
```

The command used to import multiple rasters was very similar to the one we used to import a single file. The difference was a filename mask used in place of a filename: `gray_50m_partial*.tif`. If we had used a bit more griddy pattern such as `*.tif`, all the TIF files present in a directory would be imported.

When processing multiple files, one can pipe the output to psql without the connection info specified as psql params, but in such a case, equivalent environment variables will have to be set (on Windows, use the `set` command, and on Linux, `export`):

```
set PGPORT=5434
set PGHOST=localhost
set PGUSER=postgres
set PGPASSWORD=somepass
set PGDATABASE=mastering_postgis
raster2pgsql -s 4326 -C -l 2,4 -I -F -t 2700x2700 gray_50m_partial*.tif
data_import.gray_50m_partial | psql
```

Importing data with pgrestore

Just to make the data import complete, it is worth mentioning the restore command. After all, it is not very an uncommon scenario to receive some data in the form of a database, schema, or even a single table backup.

For this scenario, let's create a backup of one of the tables imported before:

```
pg_dump -h localhost -p 5434 -U postgres -t
data_import.earthquakes_subset_with_geom -c -F c -v -b -f
earthquakes_subset_with_geom.backup mastering_postgis
```

Since there was a `-v` option specified, you should get a similarly verbose output:

```
pg_dump: reading schemas
pg_dump: reading user-defined tables
pg_dump: reading extensions
pg_dump: reading user-defined functions
pg_dump: reading user-defined types
pg_dump: reading procedural languages
pg_dump: reading user-defined aggregate functions
pg_dump: reading user-defined operators
pg_dump: reading user-defined operator classes
pg_dump: reading user-defined operator families
pg_dump: reading user-defined text search parsers
pg_dump: reading user-defined text search templates
```

```
pg_dump: reading user-defined text search dictionaries
pg_dump: reading user-defined text search configurations
pg_dump: reading user-defined foreign-data wrappers
pg_dump: reading user-defined foreign servers
pg_dump: reading default privileges
pg_dump: reading user-defined collations
pg_dump: reading user-defined conversions
pg_dump: reading type casts
pg_dump: reading transforms
pg_dump: reading table inheritance information
pg_dump: reading event triggers
pg_dump: finding extension members
pg_dump: finding inheritance relationships
pg_dump: reading column info for interesting tables
pg_dump: finding the columns and types of table
"data_import.earthquakes_subset_with_geom"
pg_dump: flagging inherited columns in subtables
pg_dump: reading indexes
pg_dump: reading constraints
pg_dump: reading triggers
pg_dump: reading rewrite rules
pg_dump: reading policies
pg_dump: reading row security enabled for table
"data_import.earthquakes_subset_with_geom"
pg_dump: reading policies for table
"data_import.earthquakes_subset_with_geom"
pg_dump: reading large objects
pg_dump: reading dependency data
pg_dump: saving encoding = UTF8
pg_dump: saving standard_conforming_strings = on
pg_dump: dumping contents of table
"data_import.earthquakes_subset_with_geom"
```

Having backed up our table, let's drop the original one:

```
DROP TABLE data_import.earthquakes_subset_with_geom;
```

And see if we can restore it:

```
pg_restore -h localhost -p 5434 -U postgres -v -d mastering_postgis
earthquakes_subset_with_geom.backup
```

You should see a similar output:

```
pg_restore: connecting to database for restore
pg_restore: creating TABLE "data_import.earthquakes_subset_with_geom"
pg_restore: processing data for table
"data_import.earthquakes_subset_with_geom"
pg_restore: setting owner and privileges for TABLE
"data_import.earthquakes_subset_with_geom"
pg_restore: setting owner and privileges for TABLE DATA
"data_import.earthquakes_subset_with_geom"
```

At this stage, we have successfully imported data by using the PostgreSQL backup / restore facilities.

 If you happen to get some errors on the pg_dump version, do make sure you're using the one appropriate for the DB you are exporting from. You can find it in the bin folder of the PostgreSQL directory.

Summary

There are many different ways of importing data into a PostGIS database. It is more than likely that you will not have to use all of them all the time but rather have your preferred ways of loading the data in your standard routines.

It is worth knowing different tools, though considered as different scenarios, may have to be addressed with special care. When you add scripting to the stack (see the chapter on ETL), you are equipped with some simple yet very powerful tools that may constitute a decent ETL toolbox.

Having fed our database with data, next, we'll have a look at spatial data analysis and find our way through a rich function set of PostGIS.

2
Spatial Data Analysis

So far, we have learned how to store geospatial data in a PostGIS database. In fact, any database management system can do that; spatial information can be encoded and stored in an ordinary DBMS-friendly format, be it blob, byte array, or text-based exchange format. What makes the spatial database special (and the PostGIS extension worth installing) is the rich toolset designed for analyzing, transforming, validating, querying, and extracting metrics from spatial information. In this chapter, we will learn how to harness the power of PostGIS spatial functions to gain meaningful insights from geodata. We will focus on the following topics:

- Composing and decomposing geometries
- Spatial measurement
- Geometry bounding boxes
- Geometry simplification
- Geometry validation
- Intersecting geometries
- Nearest feature queries

The example queries used in this chapter mostly use geometries created by hand, using geometry composition functions. For a few examples, the OSM data in `osm2pgsql` schema and Natural Earth data imported in `Chapter 1`, *Importing Spatial Data*, are used, but only the column and table names are specific; the queries can be adjusted to any other dataset, provided that the geometry type is the same.

Composing and decomposing geometries

One can think of PostGIS geometries as building blocks. The smallest unit, the point, consists of a tuple (in the most common scenario of 2D geometries, a pair) of coordinates. Points can be then used as independent units or arranged into more complex shapes: MultiPoints and LineStrings. LineStrings can constitute a MultiLineString. Closed LineStrings can be treated as rings and form a Polygon. Finally, multiple polygons may form a MultiPolygon. PostGIS is equipped with functions for coupling and decoupling those geometric building blocks, which will be outlined in this section.

Composition and decomposition functions have different names, but in general they follow a similar pattern: composition is done by supplying an array of components or using a PostgreSQL aggregation, and decomposition is done by extracting individual components by their index or by exploding a geometry into multiple rows using a set-returning function. For details of each geometry type, read on.

Creating points

This is a very important function, since tabular data with coordinates stored in separate columns is quite abundant. Examples include geocoded address lists, survey results, and telemetry measurements. Such data can be converted to geometry using the ST_MakePoint function:

```
SELECT ST_MakePoint(391390,5817855);
```

Given two numbers, it creates a point geometry. To make it clear what the location really is, and make the created geometry suitable for spatial analysis, an SRID must be given. This can be accomplished using a ST_SetSRID function. So the complete example will be:

```
SELECT ST_SetSRID(ST_MakePoint(391390,5817855),32633);
```

The number 32633 is an SRID for UTM coordinates zone 33 north, and in this case, the coordinate pair 391390,5817855 denotes the Hallesches Tor in Berlin.

When the two dimensions aren't enough, a third argument can be added with a Z-coordinate value:

```
SELECT ST_SetSRID(ST_MakePoint(334216.6,5077675.3,4810),32632);
```

This will output a 3D point geometry for Mont Blanc peak in the UTM zone 32 north coordinate system.

For geometries with both Z-coordinates and M-coordinates (M for measure, linear referencing) there are third and fourth arguments, the fourth meaning M-value:

```
SELECT ST_SetSRID(ST_MakePoint(503612.6,5789004.9,89.5,4.408),32634);
```

This will output a 4D point geometry for Warsaw East train station, located at `503612.6,5789004.9` UTM 34N, `89.5` meters above mean sea level and `4.408` kilometers from the beginning of the line at Central Station.

When a point has an `M` value but no `Z`, there is a special `ST_MakePointM` function:

```
SELECT ST_SetSRID(ST_MakePointM(503612.6,5789004.9,4.408),32634);
```

It takes three arguments: `X`, `Y`, and `M` values.

Extracting coordinates from points

Any coordinate (that is, an `X`, `Y`, `Z` or `M` value) can be extracted from a geometry into a human-readable numeric format. The `ST_X`, `SY_Y`, and `ST_Z` functions are suited for that purpose. For example, to extract the `X` and `Y` coordinates of POIs into separate numeric columns, you can use the following:

```
SELECT id, name, ST_X(way) AS x_coord, ST_Y(way) as y_coord FROM
planet_osm_point LIMIT 10;
```

Composing and decomposing Multi-geometries

There are two functions designed for creating Multi-geometries:

- The first one, `ST_Multi`, is used for converting a single geometry into a single-part Multi-geometry:

  ```
  SELECT ST_AsText(ST_Multi(ST_MakePoint(391390,5817855)))
  MULTIPOINT((391390,5817855))
  ```

  This is useful in situations where the database design enforces a uniform geometry type across the whole table, and single geometries cannot be stored alongside Multi-geometries.

  To create a multi-part Multi-geometry, another function called `ST_Collect` must be used.

The simplest use case is to merge two single geometries into a single Multi-geometry. For that, `ST_Collect` can accept two arguments of the geometry type:

```
SELECT
ST_Collect(ST_MakePoint(20,50),ST_MakePoint(19.95,49.98));
```

For more complex MultiPoints, there are two possibilities. The first one is to pass the PostgreSQL `ARRAY` of geometries as the argument to `ST_Collect`:

```
SELECT
ST_Collect(ARRAY[ST_MakePoint(20,50),ST_MakePoint(19.95,49.98),
ST_MakePoint(19.90,49.96)]);
```

- The second option is to use `ST_Collect` as an aggregate function, just like `SUM` or `AVG` are used for numbers. For example, to collect POI groups into MultiPoints, you can do the following:

```
SELECT amenity, ST_Collect(way) FROM planet_osm_point GROUP BY
amenity;
```

When using `ST_Collect` as an aggregate, the standard rules of using aggregate functions in PostgreSQL apply: All other columns must be used in another aggregate (`SUM`, `AVG`, `string_agg`, `array_agg`, and so on) or in a `GROUP BY` clause. Failure to do so will result in an error:

```
SELECT tourism, amenity, ST_Collect(way) FROM planet_osm_point
GROUP BY amenity;

ERROR: column planet_osm_point.tourism must appear in the
GROUP BY clause or be used in an aggregate function.
```

Multi-geometry decomposition

Multi-geometries can be broken down into single parts. There are two ways to do this.

- The first is to extract components one part at a time, using the `ST_GeometryN` function.

 In computer programming in general, indexes are 0 based, so the first list element will have an index of 0. In PostgreSQL, however, the convention is different and indexes are 1 based. This means that the first element has an index of 1.

For example, extracting the second part from a MultiPoint created by hand can be done using the following:

```
SELECT
ST_AsText(ST_GeometryN(ST_Collect(ARRAY[ST_MakePoint(20,50),
ST_MakePoint(19.95,49.98), ST_MakePoint(19.90,49.96)]),2));

  st_astext
--------------------
 POINT(19.95 49.98)
```

The number of Multi-geometry parts can be revealed using the ST_NumGeometries function:

```
SELECT
ST_NumGeometries(ST_Collect(ARRAY(ST_MakePoint(20,50),
ST_MakePoint(19.95,49.98), ST_MakePoint(19.90,49.96))));

 3
```

- The second option is to expand a Multi-geometry into a set of rows using the ST_Dump function. This returns a special compound type, called **geometry dump**. The geometry dump type consists of two parts: the path, which denotes the position of a component within the Multi-geometry, and the geom being the actual component geometry:

```
SELECT
ST_Dump(ST_Collect(ARRAY[ST_MakePoint(20,50),
ST_MakePoint(19.95,49.98), ST_MakePoint(19.90,49.96)]));
```

This will result in the following:

```
st_dump
------------------------------------------------------
({1},01010000000000000000003440000000000004940)
({2},01010000003333333333F333403D0AD7A370FD4840)
({3},01010000006666666666E633407B14AE47E1FA4840)
(3 rows)
```

To extract actual geometries from the geometry dump, the ST_Dump function should be wrapped in parentheses, and its geom part referenced:

```
SELECT
(ST_Dump(ST_Collect(ARRAY[ST_MakePoint(20,50),
ST_MakePoint(19.95,49.98), ST_MakePoint(19.90,49.96)]))).geom;
                    geom
------------------------------------------------
```

```
01010000000000000000000003440000000000000004940
01010000003333333333F333403D0AD7A370FD4840
01010000006666666666E633407B14AE47E1FA4840
```

Now each of the MultiPoint's components are accessible as a single-part geometry, which can be inserted into another table or used in another function.

Composing and decomposing LineStrings

LineStrings are ordered collections of points used for representing linear spatial features such as roads, railways, rivers, or power lines. PostGIS has functions for creating them, given a set of points, and decomposing them into sets of points.

LineString composition

The rules for composing LineStrings are similar to those governing MultiPoint composition: Two points, an array of points, or a set of aggregated rows can be supplied. The PostGIS function designed for creating LineStrings is called ST_MakeLine. Point geometries required for composition can be already stored in a table, or created from raw coordinates using the ST_MakePoint function.

The following example will create a straight line connecting two points:

```
SELECT ST_MakeLine(ST_MakePoint(20,50),ST_MakePoint(19.95,49.98));
```

When using raw coordinates, the output geometry will have an unknown SRID, so the complete example will be as follows:

```
SELECT
ST_SetSRID(ST_MakeLine(ST_MakePoint(20,50),ST_MakePoint(19.95,49.98)),4326)
;
```

For three or more points and raw coordinates, the ARRAY argument can be used:

```
SELECT ST_MakeLine(ARRAY[ST_MakePoint(20,50),ST_MakePoint(19.95,49.98),
ST_MakePoint(19.90,49.96)]);
```

And finally, the aggregate variant. This is especially useful when dealing with a series of points from GPS tracking devices:

```
SELECT ST_MakeLine(gpx.geom ORDER BY time) AS geom
FROM gpx
GROUP BY 1;
```

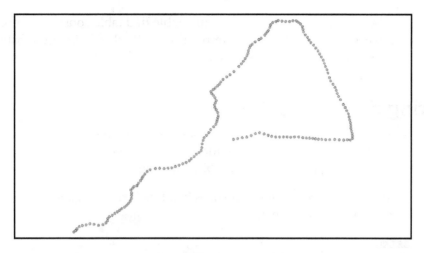

A set of GPS measurement points.

The image we just saw contains a visualization of discrete GPS points. Each one has a **time** attribute that can be used for sorting, so a LineString can be composed without issues.

In the next step, the sorted points are composed into a single geometry of the LineString type:

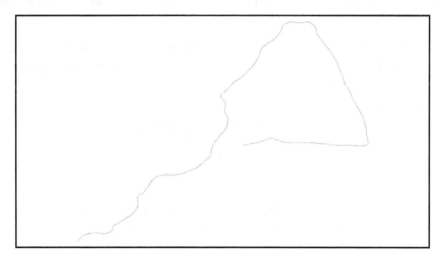

A LineString created from a set of points.

The `GROUP BY 1` is used to aggregate all rows in a table (if a table consists of multiple tracks and they have a unique ID, a real column with this ID should be used instead), and the `ORDER BY` timestamp clause ensures the correct order of points to create a valid line.

LineString decomposition

This process is also similar to MultiPoint decomposition. The points can be extracted one at a time or dumped into a geometry dump. Aside from these generic methods, there are also special functions for finding the start or end point of a line.

1. First, let's find the count of the vertices in a LineString. A generic `ST_NPoints` function can be used for that:

   ```
   SELECT
   ST_NPoints(ST_MakeLine(ST_MakePoint(20,50),
   ST_MakePoint(19.95,49.98)));
   ```

   ```
   2
   ```

   The `ST_NPoints` works for any geometry type--not just for LineStrings, but also LinearRings, polygons, their multi-variants, and even for points, in which case it will return 1. There is also a `ST_NumPoints` function, which works for LineStrings only.

   The individual points are extracted from a LineString by the `ST_PointN` function, which is not unlike `ST_GeometryN` used for Multi-geometry decomposition.

   ```
   SELECT
   ST_AsText(ST_PointN(ST_MakeLine(ARRAY[
   ST_MakePoint(20,50),ST_MakePoint(19.95,49.98),
   ST_MakePoint(19.90,49.96)]),2));
   ```

   ```
   POINT(19.95 49.98)
   ```

2. Dumping the points into a set of rows is also similar to dumping a Multi-geometry. The only difference is that the function used is named `ST_DumpPoints`. The geometry dump compound type is a direct output from this function:

   ```
   SELECT
   ST_DumpPoints(ST_MakeLine(ARRAY[
   ST_MakePoint(20,50),ST_MakePoint(19.95,49.98),
   ST_MakePoint(19.90,49.96)]));
   ```

```
st_dumppoints
-------------------------------------------------------
({1},0101000000000000000000003440000000000000004940)
({2},01010000003333333333F333403D0AD7A370FD4840)
({3},01010000006666666666E633407B14AE47E1FA4840)
```

3. And for accessing the vertices' geometry, we can do the following:

```
SELECT
(ST_DumpPoints(ST_MakeLine(ARRAY[
ST_MakePoint(20,50),ST_MakePoint(19.95,49.98),
ST_MakePoint(19.90,49.96)])))).geom;
                        geom
-------------------------------------------------
0101000000000000000000003440000000000000004940
01010000003333333333F333403D0AD7A370FD4840
01010000006666666666E633407B14AE47E1FA4840
```

4. Finally, there are specialized methods for finding the first and last vertices of LineStrings. They are named ST_StartPoint and ST_EndPoint:

```
SELECT
ST_AsText(ST_StartPoint(ST_MakeLine(ARRAY[
ST_MakePoint(20,50),ST_MakePoint(19.95,49.98),
ST_MakePoint(19.90,49.96)])));

POINT(20 50)

SELECT
ST_AsText(ST_EndPoint(ST_MakeLine(ARRAY[
ST_MakePoint(20,50),ST_MakePoint(19.95,49.98),
ST_MakePoint(19.90,49.96)])));

POINT(19.9 49.96)
```

Composing and decomposing polygons

Polygons in PostGIS are made of rings. A ring is a closed LineString-that is, the first and last point have the same coordinates. A polygon must have only one outline (also called a shell, or exterior ring) and can have zero or more interior rings (holes), which must be contained within the outer ring. The winding order of vertices is irrelevant (in contrast to some other GIS formats, most notably the Shapefile format). This means that if a polygon feature has more than one exterior ring, it must be saved as a MultiPolygon.

Polygon composition

A function for composing polygons is not-so-surprisingly named `ST_MakePolygon`. For a solid polygon without holes, it accepts one argument-a closed LineString.

1. For example, a triangle-shaped polygon can be created using the following query:

```
SELECT
ST_MakePolygon(ST_MakeLine(ARRAY[ST_MakePoint(20,50),
ST_MakePoint(19.95,49.98),
ST_MakePoint(19.90,49.90),ST_MakePoint(20,50)]));
```

2. An attempt to create a polygon from fewer than four points will result in an error:

```
SELECT
ST_MakePolygon(ST_MakeLine(ARRAY[ST_MakePoint(20,50),
ST_MakePoint(19.95,49.98), ST_MakePoint(19.90,49.96)]));
ERROR:  lwpoly_from_lwlines: shell must have at least 4 points
```

And if a ring is not closed:

```
SELECT
ST_MakePolygon(ST_MakeLine(ARRAY[ST_MakePoint(20,50),
ST_MakePoint(19.95,49.98),
ST_MakePoint(19.90,49.96),ST_MakePoint(20,50),
ST_MakePoint(20.01,50.01)]));
ERROR:  lwpoly_from_lwlines: shell must be closed
```

When creating simple shapes using coordinates typed by hand, it's rather obvious if a ring is closed or not. But what if there are hundreds or thousands of vertices? Luckily, PostGIS has a function for finding out whether a LineString is closed or not--it's called `ST_IsClosed`:

```
SELECT
ST_IsClosed(ST_MakeLine(ARRAY[ST_MakePoint(20,50),
ST_MakePoint(19.95,49.98),
ST_MakePoint(19.90,49.96),ST_MakePoint(20,50),
ST_MakePoint(20.01,50.01)]));
 st_isclosed
-------------
 f
```

```
SELECT
ST_IsClosed(ST_MakeLine(ARRAY[ST_MakePoint(20,50),
ST_MakePoint(19.95,49.98),
ST_MakePoint(19.90,49.96),ST_MakePoint(20,50),
ST_MakePoint(20,50)]));
```

```
st_isclosed
--------------
t
```

When using Boolean-returning functions in the PostgreSQL console, f stands for false, and t for true.

PostGIS can also snap the ends of a ring to create a polygon from an unclosed LineString. Just replace ST_MakePolygon with ST_Polygonize:

```
SELECT
ST_Polygonize(ST_MakeLine(ARRAY[ST_MakePoint(20,50),
ST_MakePoint(19.95,49.98),
ST_MakePoint(19.90,49.96),ST_MakePoint(20,50),
ST_MakePoint(20.01,50.01)]));
 st_polygonize
---------------------
 010700000000000000
(1 row)
```

Beware though--this is an automated process, and for some shapes it can lead to undesirable effects, like weirdly-shaped MultiPolygons.

3. For constructing a polygon with holes, the interior ring(s) should be supplied as an ARRAY:

```
SELECT ST_MakePolygon(
ST_MakeLine(ARRAY[ST_MakePoint(20,50),ST_MakePoint(19.95,49.98),
ST_MakePoint(19.90,49.90),ST_MakePoint(20,50)]),
ARRAY[ST_MakeLine(ARRAY[ST_MakePoint(19.95,49.97),ST_MakePoint
(19.943,49.963),
ST_MakePoint(19.955,49.965),ST_MakePoint(19.95,49.97)])]
);
```

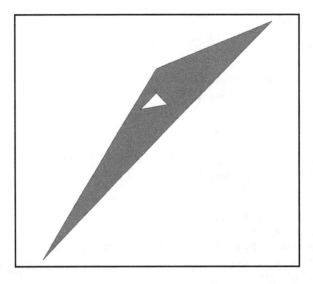

An example polygon with exterior and interior rings created with ST_MakePolygon.

Polygon decomposition

There are two levels of polygon geometry decomposition: they can be broken down either into LineStrings or into points. Decomposition into points is very similar to decomposition of a LineString: the ST_DumpPoints is used:

```
SELECT ST_DumpPoints(
    ST_MakePolygon(
        ST_MakeLine(ARRAY[ST_MakePoint(20,50),ST_MakePoint(19.95,49.98),
ST_MakePoint(19.90,49.90),ST_MakePoint(20,50)]),
    ARRAY[ST_MakeLine(ARRAY[ST_MakePoint(19.95,49.97),ST_MakePoint(19.943,49.96
3), ST_MakePoint(19.955,49.965),ST_MakePoint(19.95,49.97)])]
    )
);
```

```
                    st_dumppoints
---------------------------------------------------------
 ("{1,1}",0101000000000000000000034400000000000004940)
 ("{1,2}",01010000003333333333F333403D0AD7A370FD4840)
 ("{1,3}",01010000006666666666E6334033333333333F34840)
 ("{1,4}",0101000000000000000000034400000000000004940)
 ("{2,1}",01010000003333333333F3334405C8FC2F528FC4840)
 ("{2,2}",0101000000C520B07268F133402506819543FB4840)
 ("{2,3}",010100000014AE47E17AF43340EC51B81E85FB4840)
 ("{2,4}",01010000003333333333F3334405C8FC2F528FC4840)
```

There's a slight difference in the geometry dump format: the path array now consists of two numbers. The first one is the index of the ring (1 being the exterior ring, and the following numbers the subsequent interior rings), and the second is the index a of point in the ring.

For extracting whole rings, PostGIS offers the following possibilities:

- Dumping the rings into a set of rows. For that, ST_DumpRings is used:

```
SELECT ST_DumpRings(
ST_MakePolygon(
ST_MakeLine(ARRAY[ST_MakePoint(20,50),ST_MakePoint(19.95,49.98),
ST_MakePoint(19.90,49.90),ST_MakePoint(20,50)]),

ARRAY[ST_MakeLine(ARRAY[ST_MakePoint(19.95,49.97),ST_MakePoint
(19.943,49.963),
ST_MakePoint(19.955,49.965),ST_MakePoint(19.95,49.97)])]
  )
  );
```

```
        st_dumprings
-----------------------------------
({1},0103000000010000004000000(...)
```

- Extracting the exterior ring:

```
SELECT ST_ExteriorRing(
ST_MakePolygon(
ST_MakeLine(ARRAY[ST_MakePoint(20,50),ST_MakePoint(19.95,49.98),
ST_MakePoint(19.90,49.90),ST_MakePoint(20,50)]),

ARRAY[ST_MakeLine(ARRAY[ST_MakePoint(19.95,49.97),
ST_MakePoint(19.943,49.963),ST_MakePoint(19.955,49.965),
ST_MakePoint(19.95,49.97)])]
  )
  );
```

```
        st_exteriorring
---------------------------
01020000000400000000(...)
```

- Extracting the interior rings based on their index:

```
SELECT ST_InteriorRingN(
ST_MakePolygon(
ST_MakeLine(ARRAY[ST_MakePoint(20,50),ST_MakePoint(19.95,49.98),
ST_MakePoint(19.90,49.90),ST_MakePoint(20,50)]),

ARRAY[ST_MakeLine(ARRAY[ST_MakePoint(19.95,49.97),
ST_MakePoint(19.943,49.963),
ST_MakePoint(19.955,49.965),ST_MakePoint(19.95,49.97)])]
 ), 1
);
```

```
      st_interiorringn
---------------------------
0102000000040000000333(...)
```

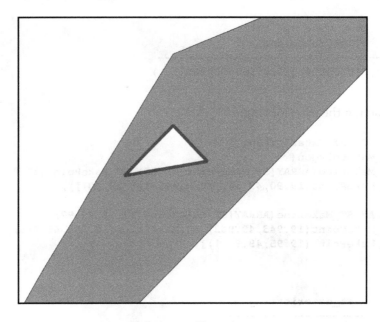

Interior ring extracted from a polygon.

Spatial measurement

Before the advent of GIS, spatial measurement was quite a tedious task. On printed (or drawn) maps, this usually meant aligning a piece of string or wire on a map carefully and then measuring it with a ruler. Another possibility was to use a specialized tool, called an **opisometer** (**curvimeter**). The whole process was error prone and subject to a number of factors, including the stretching or shrinking of the map material. For surveyed coordinates, measurement meant doing lots of math. Things were even worse for geodesic (latitude-longitude) coordinates, where complicated formulas had to be used.

Now, with spatial databases, it's possible to get spatial measurements in milliseconds. This doesn't mean that one can blindly push the button, though. The details will be explained in this section.

General warning - mind the SRID!

Measurement functions come in two variants: planimetric and geodesic. The former operate on a flat surface, while the latter operate on a curved surface (sphere or spheroid). **Planimetric** functions are robust and computationally cheap (and thus fast to execute), but will yield accurate results if and only if the data is stored in a suitable Cartesian coordinate system. Examples include UTM, State Plane for the US, National Grid for the US and UK, DHDN for Germany, PUWG for Poland, S-JTSK for the Czech Republic, and so on. Geodesic functions, on the other hand, are suited for latitude-longitude coordinates, and consume more resources because of their more complicated formulas, but they work globally.

Trying to use planimetric functions on geodesic coordinates will not result in an error or warning, but will give meaningless results instead. For instance, the distance between Brasilia and Rio de Janeiro in Brazil or Kristiansand in Norway and Espoo in Finland is about 990 kilometers, but 9.1 decimal degrees in the first case and 16.79 decimal degrees in the second. This is because Earth is not flat, and the meridians converge towards the poles. The distance between two meridians is not uniform across the globe, so decimal degrees cannot be used as a reliable unit of measurement.

Attention should be paid to global flat coordinate systems, too. Web Mercator is probably the most notable example, as databases used primarily for web map tile rendering use it. Using planimetric measures in this coordinate system will result in exaggerated results--the closer to the poles, the bigger the distortion. For example, the area of Spain calculated using geodesic function is 507,000 square kilometers, but with Web Mercator it is 873,000 square kilometers (72 percent bigger), and the measurement of Sweden's area is 446,000 square kilometers with the geodesic function and more than 2 million square kilometers with Web Mercator (480 percent distortion!).

Measuring distances between two geometries

In order to measure the distance between two geometries, the ST_Distance function should be used. It accepts two arguments of the geometry (or geography) type, and when using the more common geometry type, it returns a floating point number indicating the distance in the same units as the geometries' SRID. So for UTM the result will be given in meters, for State Plane in feet, and for WGS84 in decimal degrees (which, as I mentioned before, is plainly useless). Both geometries need to have the same SRID or an error will be thrown. For example, we can use the following to calculate the distance between Potsdam and Berlin:

```
SELECT ST_Distance(
    ST_SetSRID(ST_MakePoint(367201,5817855),32633),
    ST_SetSRID(ST_MakePoint(391390,5807271),32633)
)

    st_distance
------------------
  26403.1963405948
```

But what if the geometries are in a latitude-longitude coordinate system? This is where geodesic measurement comes into play. When the features are in the geography type, all the user has to do is nothing; PostGIS will calculate a geodesic length and return the result in meters. If the geometry type is used, it can be cast to geography:

```
SELECT
ST_Distance(ST_MakePoint(20,50)::geography,ST_MakePoint(21,51)::geography);
```

On modern hardware, this will run fast despite the complexity of math involved. But for large number of features it can slow down things anyway. When time and/or processing power is at a premium, and accuracy can be sacrificed - ,PostGIS allows for using a simpler Earth model, a sphere instead of spheroid. For this, an optional second argument `use_spheroid = FALSE` has to be supplied. As you might remember from geography class, the Earth is not a perfect sphere, but it's slightly flattened. When using a spherical model for calculation, the accuracy of measurement will decrease. Decreased by how much, you ask? This depends on the latitude and distance between features, but here are some examples.

For two landmarks in Berlin, the spheroidal distance is 2157.5 meters and the spherical distance is 2155 meters--a 0.116 percent difference.

For the cities of Brasilia and Rio de Janeiro in Brazil, the spheroidal distance is 988.02 kilometers and the spherical distance is 990.31 kilometers--a 0.232 percent difference.

For the terminal stations of the Trans-Siberian Railway, Moscow and Vladivostok, the spheroidal distance is 6430.7 kilometers and the spherical distance is 6412 kilometers - a 0.44 percent difference.

The query will look like this:

```
SELECT
ST_Distance(ST_MakePoint(20,50)::geography,ST_MakePoint(21,51)::geography,
FALSE);
```

The speedup for simple points is about 10 percent, but it will be bigger as the geometries become more complex.

 Before the introduction of the geography type, specialized functions named `ST_DistanceSphere` and `ST_DistanceSpheroid` had to be used for latitude-longitude coordinates.

Measuring the length, area, and perimeter of geometries

PostGIS can not only measure distances between geometries, but also the dimensions of line and polygon geometries.

Line length

For lines, the length is calculated by the ST_Length function. It takes one argument, the input geometry, which can be a LineString or MultiLineString. When the argument is of the geometry type, the planimetric calculation will take place and the output length will be given in units of the input geometry's coordinate system.

Here is an example of a three-point LineString:

```
SELECT ST_Length(
ST_MakeLine(ARRAY[ST_MakePoint(391390,5817855),ST_MakePoint(391490,5817955)
, ST_MakePoint(391590,5818055)])
);

    st_length
-------------------
 282.842712474619
```

For latitude-longitude coordinates, either the geography type, a type cast, or a ST_Length_Spheroid function should be used. Here is an example of a type cast being used:

```
SELECT ST_Length(
    ST_MakeLine(ARRAY[ST_MakePoint(20,50),ST_MakePoint(19.95,49.98),
ST_MakePoint(19.90,49.96)])::geography
);

    st_length
-------------------
 8440.40084752485
```

The result will be given in meters.

For polygons, perimeter and area calculations are available.

Polygon perimeter

The perimeter is computed using the ST_Perimeter function. Its usage is identical to ST_Length, the only difference being that the argument must be a polygon or MultiPolygon:

```
SELECT ST_Perimeter(ST_GeomFromText('POLYGON((391390 5817855,391490
5817955,391590 5818055, 319590 5817855,391390 5817855))', 32633))

    st_perimeter
-------------------
```

```
144083.120489717
```

Again, for latitude-longitude coordinates, the geography type should be used. Please note that in contrast to distance and length computations, there's no specialized function for the geodesic calculation of the perimeter of latitude-longitude geometry. Here is an example, using the triangle from previous section:

```
SELECT ST_Perimeter(
ST_MakePolygon(ST_MakeLine(ARRAY[ST_MakePoint(20,50),ST_MakePoint(19.95,49.
98), ST_MakePoint(19.90,49.90),ST_MakePoint(20,50)])))::geography
);

    st_perimeter
--------------------
27051.7310753665
```

When a polygon has holes, the perimeter of the interior rings will be added to the exterior rings' perimeter:

```
SELECT ST_Perimeter(ST_MakePolygon(
    ST_MakeLine(ARRAY[ST_MakePoint(20,50),ST_MakePoint(19.95,49.98),
ST_MakePoint(19.90,49.90),ST_MakePoint(20,50)]),
ARRAY[ST_MakeLine(ARRAY[ST_MakePoint(19.95,49.97),ST_MakePoint(19.943,49.96
3), ST_MakePoint(19.955,49.965),ST_MakePoint(19.95,49.97)])]
)::geography);

    st_perimeter
--------------------
29529.3133397193
```

Polygon area

The area of a polygon or MultiPolygon geometry is computed with the `ST_Area` function. The same rules that we used with the perimeter calculations apply: when the argument is of the geometry type, the planimetric variant will be used, and with the geography type, the geodesic variant will be used:

```
SELECT ST_Area(ST_GeomFromText('POLYGON((391390 5817855,391490
5817955,391590 5818055, 319590 5817855,391390 5817855))', 32633));
    st_area
----------
7180000
```

The argument is of the geometry type, and the SRID is 32633, which leads us to UTM 33N. The result is given in square meters:

```
SELECT ST_Area(
ST_MakePolygon(ST_MakeLine(ARRAY[ST_MakePoint(20,50),ST_MakePoint(19.95,49.
98), ST_MakePoint(19.90,49.90),ST_MakePoint(20,50)]))::geography
);

     st_area
-------------------
 11949254.7990036
```

The argument is of the geography type, so the result will be computed geodesically with the units as square meters.

Finally, for polygons with holes, the interior rings' area is subtracted from the exterior rings' area:

```
SELECT ST_Area(ST_MakePolygon(
    ST_MakeLine(ARRAY[ST_MakePoint(20,50),ST_MakePoint(19.95,49.98),
ST_MakePoint(19.90,49.90),ST_MakePoint(20,50)]),
ARRAY[ST_MakeLine(ARRAY[ST_MakePoint(19.95,49.97),ST_MakePoint(19.943,49.96
3), ST_MakePoint(19.955,49.965),ST_MakePoint(19.95,49.97)])]
)::geography);

     st_area
-------------------
 11669942.4506721
```

Geometry bounding boxes

A bounding box, often abbreviated into BBOX, is a list of the extreme coordinates of a geometry. Bounding boxes play a big role in spatial queries, as they allow for fast coarse computations; If two geometries' BBOXes do not intersect, there's no point in wasting CPU cycles for intersecting them precisely. In PostGIS, bounding boxes are computed and cached internally. There are specialized data types for bounding boxes: box2d, which can be visualized as a rectangle, and box3d, which forms, well, a box.

Accessing bounding boxes

A bounding box of a geometry can be accessed in two ways. The first is to create a `Box2D` type from a geometry (or a set of them):

```
SELECT ST_Extent(
ST_GeomFromText('POLYGON((391390 5817855,391490 5817955,391590 5818055,
319590 5817855,391390 5817855))', 32633)
);
            st_extent
----------------------------------------
 BOX(319590 5817855,391590 5818055)
```

Then to create one for a set of geometries:

```
SELECT ST_Extent(geom) FROM sometable;
```

`ST_Extent` is an aggregate function, so the usual rules of using the `GROUP BY` clause apply.

The second method is to access individual BBOX elements separately. For that, `ST_XMin`, `ST_YMin`, `ST_XMax`, and `ST_YMax` (and `ST_ZMax` for 3D geometries) are provided:

```
SELECT ST_XMin(
ST_GeomFromText('POLYGON((391390 5817855,391490 5817955,391590 5818055,
319590 5817855,391390 5817855))', 32633)
);

 st_xmin
---------
 319590
```

For geometries already stored in a table, we can use the following:

```
SELECT ST_XMin(geom), ST_YMin(geom), ST_XMax(geom), ST_YMax(geom) FROM
sometable;
```

This function is not an aggregate, so for computing individual extreme coordinates for sets of geometries, they need to be aggregated first using the `ST_Collect` function that we learned about before.

Creating bounding boxes

Like regular geometries, a bounding box can be created from its textual representation:

```
SELECT 'BOX(319590 5817855,391590 5818055)'::box2d;
```

It can also be created using a function, ST_MakeBox2D, which accepts two points:

```
SELECT ST_MakeBox2D(
    ST_MakePoint(319590,5817855),
    ST_MakePoint(391590,5818055)
);
```

The first argument is a lower-left point, with coordinates of (minX,minY), and the second is an upper-right point, with coordinates of (maxX,maxY).

Using bounding boxes in spatial queries

The ability to create a bounding box is useful for doing filter by BBOX queries. The coordinate systems of the data and the BBOX must match, so any BBOX creation function must be wrapped in the ST_SetSRID function. For that kind of query, the && operator, meaning intersects BBOX, is used. For example:

```
SELECT * FROM planet_osm_point WHERE way && ST_SetSRID('BOX(-500
6705000,1000 6706000)'::box2d,900913);
```

This will return all the rows from the planet_osm_point table whose geometries fall completely or partially inside the defined BBOX. Note that this returns the complete geometries that share even the tiniest part with a BBOX. Clipping geometries to BBOX will be discussed in the next chapter.

Geometry simplification

In surveying, we strive for the highest accuracy and precision possible. However, this is not always the case in mapmaking. Too-detailed geometries on small scale (zoomed-out) maps are bad for human perception; edges or lines appear jagged, and cause an unnecessary burden on the computer in trying to render them. For that reason, cartographers generalize the geometries used for mapmaking. There are a few algorithms designed for automated generalization. PostGIS provides an implementation of a widely used Douglas-Peucker algorithm in a ST_Simplify function.

The function accepts two arguments, the first being a geometry to be simplified, and the second being a tolerance parameter defining how aggressive the simplification can be. Tolerance is given in the same units as the geometry's coordinate system, and the bigger it is, the more simplified the output geometry becomes.

Here are some examples of simplification:

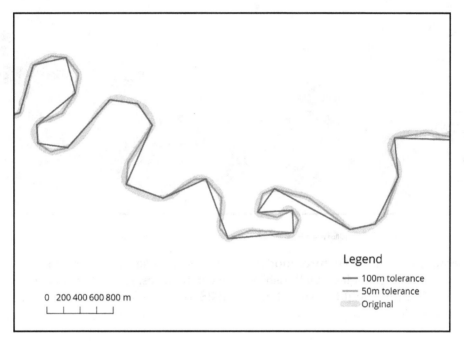

River meanders simplified with 50 (green) and 100 (red) meter tolerance. The original geometry is blue.

The problem with the ST_Simplify function is that it can break geometries that introduce self-intersections, for example (more on geometry validity in the next section). To address this, a safer variant called ST_SimplifyPreserveTopology was introduced. This function will do its best to avoid creating invalid geometries.

Please note that this function maintains topology within single rows only. It can--and it will--break topological relationships between adjacent geometries, for example, land parcels or river networks:

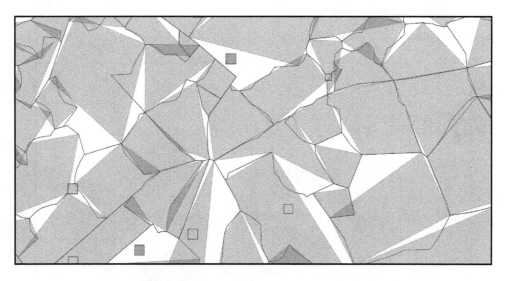

Aggressively simplified postcode polygons that have retained validity, but lost their topological relationships.

This is because the Simple Features model, which PostGIS uses, does not store any information about topological relationships between features. If topological correctness is required, the features should be stored in a PostGIS topology format (which is discussed in Chapter 8, *PostGIS Topology*) instead.

Geometry validation

Invalid geometries are a spatial analyst's nightmare. They can appear in any dataset, and can break a carefully-designed, long-running query in the middle of execution. Or even worse, a failing query might break an application's functionality. Luckily, PostGIS is equipped with the tools to find and repair them.

Simplicity and validity

In PostGIS, there are two concepts: simplicity and validity. For most spatial analyses to succeed, input geometries have to be both simple and valid. Here are some rules:

- A point must always be simple and valid
- A MultiPoint must always be valid, and simple when there are no repeated points with identical coordinates
- A LineString or MultiLineString must always be valid, and is simple if the line:
    - Does not have repeated points (with the exception of closed rings, whose first and last point are identical)
    - Does not self-intersect

Example of a non-simple line: self-intersecting autogenerated contours.

A polygon is always simple, and is valid if:

- All interior rings (if any) are properly contained within the exterior ring
- No rings cross
- The exterior ring doesn't have any **spikes** (**bayonets**)

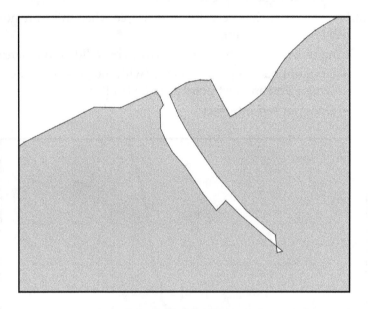

An example of an invalid polygon: self-intersecting ring.

A MultiPolygon is valid when all parts are valid and the exterior rings of all parts do not intersect.

Testing for simplicity and validity

Testing geometries for simplicity is done by the ST_IsSimple function. It accepts a geometry and returns a Boolean indicating whether a given geometry is simple. For example, a simple line looks like the following:

```
SELECT
ST_IsSimple(ST_MakeLine(ST_MakePoint(20,50),ST_MakePoint(19.95,49.98)));

 st_issimple
--------------
 t
```

The next line has a repeated point:

```
    SELECT
ST_IsSimple(ST_MakeLine(ARRAY[ST_MakePoint(20,50),ST_MakePoint(19.95,49.98)
,ST_MakePoint(19.90,49.90), ST_MakePoint(19.95,49.98)]));

 st_issimple
--------------
   f
```

So the ST_IsSimple function returns FALSE.

For geometries stored in a table, the following syntax can be used to show the non-simple ones:

```
SELECT * FROM planet_osm_line WHERE ST_IsSimple(way) = FALSE;
```

Checking for validity

There are three PostGIS functions for checking validity. The first one is the ST_IsValid function. It behaves exactly like ST_IsSimple: it accepts a geometry and returns a Boolean indicating whether the input is valid or not.

```
SELECT * FROM data_import.ne_110m_land WHERE ST_IsValid(geom) =  FALSE;
```

But what's exactly wrong with returned geometries? PostGIS can return the details of a validity check. Those can be returned in two variants: as a human readable summary or a compound type that includes the summary and location in geometry format. The first variant is named ST_IsValidReason:

```
SELECT ST_IsValidReason(geom) FROM data_import.ne_110m_land WHERE
ST_IsValid(geom) =  FALSE;
```

It returns a summary of test results:

```
                        st_isvalidreason
---------------------------------------------------------------
 Ring Self-intersection[-132.710007884431 54.0400093154234]
```

The second variant is named ST_IsValidDetail, and returns a valid_detail compound type (similar to a geometry dump) containing three fields:

- **valid**, of a Boolean type, indicating whether the geometry is valid
- **reason**, of a text type, indicating the reason for its invalidity (if any)

- **location**, of a geometry type, indicating the location at which the error was encountered

This variant is useful when making validation reports.

For example:

```
    SELECT ST_IsValidDetail(geom) FROM data_import.ne_110m_land WHERE
ST_IsValid(geom) = FALSE;
NOTICE:  Ring Self-intersection at or near point -132.71000788443121
54.040009315423447
                            st_isvaliddetail
-------------------------------------------------------------------------
 (f,"Ring Self-intersection",0101000000018717462B89660C0B8A376061F054B40)
```

The following is used to explode it into separate columns:

```
SELECT (ST_IsValidDetail(geom)).location, (ST_IsValidDetail(geom)).reason
FROM data_import.ne_110m_land WHERE ST_IsValid(geom) = FALSE;

NOTICE:  Ring Self-intersection at or near point -132.71000788443121
54.040009315423447
                    location                 |          reason
---------------------------------------------+-------------------------
 0101000000018717462B89660C0B8A376061F054B40 | Ring Self-intersection
```

Repairing geometry errors

PostGIS can try to fix validity errors automatically. A ST_MakeValid function is designed for this. It's convenient to use, but it's not a magic wand; if a geometry is an hourglass-shaped polygon, it will remain so, only converted to a MultiPolygon to ensure formal validity. Sometimes it's better to extract invalid geometries and have a closer look at them using desktop GIS software. When this is not practical, ST_MakeValid can be used as a quick remedy:

```
UPDATE data_import.ne_110m_land SET geom = ST_MakeValid(geom);
```

The ST_MakeValid will not touch geometries that are already valid, so an additional WHERE clause is not needed.

Now, look at the following query:

```
SELECT * FROM data_import.ne_110m_land WHERE ST_IsValid(geom) = FALSE;
```

This query should `return 0` rows.

There is also a **hacky** way for making geometries valid:

```
UPDATE data_import.ne_110m_land SET geom = ST_Buffer(geom,0) WHERE
ST_IsValid(geom) = FALSE;
```

Validity constraint

When the ability to run spatial queries is mission critical (for example, in a web application), it's wise to put a `CHECK` constraint on a geometry column. This will ensure that no invalid geometries will make their way into the database, so users won't be surprised by a failing query. On the other hand, an `INSERT` or `UPDATE` query with invalid geometry will fail, so this has to be correctly handled.

The constraint is created after all geometries are made valid (or no geometries have been inserted yet), using a query like this:

```
ALTER TABLE planet_osm_polygon ADD CONSTRAINT enforce_validity CHECK
(ST_IsValid(way));
```

Intersecting geometries

In PostGIS, there are two ways in which geometries can be intersected. The first way is checking whether two geometries intersect (at least one point of geometry A lies within the interior of geometry B). That's the `ST_Intersects` function. It accepts two arguments of the geometry type, and returns a Boolean.

For example, these lines intersect:

```
SELECT ST_Intersects(
    ST_MakeLine(ST_MakePoint(20,50),ST_MakePoint(21,51)),
    ST_MakeLine(ST_MakePoint(20.5,50.5), ST_MakePoint(22,52))
);
```

But these lines don't:

```
    st_intersects
---------------
 t

SELECT ST_Intersects(
    ST_MakeLine(ST_MakePoint(20,50),ST_MakePoint(21,51)),
```

```
      ST_MakeLine(ST_MakePoint(21.5,51.5), ST_MakePoint(22,52))
);

 st_intersects
----------------
  f
```

ST_Intersects can also be used for table joining. The following query will return land features that had at least one earthquake:

```
SELECT ne_110m_land.* FROM data_import.ne_110m_land JOIN
data_import.earthquakes_subset_with_geom ON
ST_Intersects(ne_110m_land.geom, earthquakes_subset_with_geom.geom);
```

The second way of intersecting features is by using the ST_Intersection function. Despite its similar name, it does a different job: it accepts two arguments of the geometry type, and returns a geometry containing the portion shared by those two features.

For example, let's compute an intersection of the British coastline with an arbitrary polygon:

```
SELECT gid,ST_Intersection(geom,
ST_SetSRID('POLYGON((-2 53, 2 53, 2 50, -2 50, -2 53))'::geometry,4326)
) FROM data_import.ne_coastline WHERE gid = 73;
```

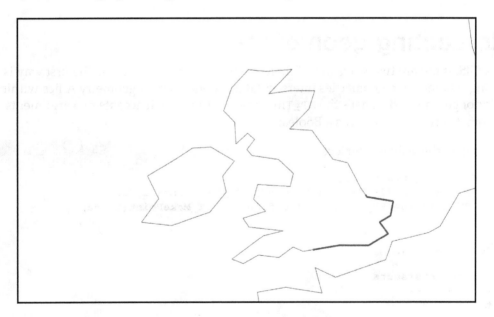

A LineString clipped with a polygon using the ST_Intersection function.

Nearest feature queries

This is a key feature for almost any location-based application: given a current location, it will return the nearest feature (or list of nearest features, ordered by distance).

The naive approach to this problem would be to query the table ordering by ST_Distance. Let's find the five earthquakes closest to San Juan:

```
SELECT * FROM data_import.earthquakes_subset_with_geom
ORDER BY ST_Distance(geom::geography,
ST_SetSRID(ST_MakePoint(-66.11,18.46),4326)::geography)
LIMIT 5;
```

```
    id      |            time          | depth | mag | magtype
|                      place
|                      geom
------------+--------------------------+-------+-----+---------+------------
------------+--------------------------+-------+------------------------------
-------------------+
 pr16281009 | 2016-10-08 01:08:46.4+02 |    5 | 2.5 | Md      | 28km SE of
El Negro, Puerto Rico            |
0101000020E610000062A1D634EF6850C0E561A1D634DF3140
 pr16281009 | 2016-10-08 01:08:46.4+02 |    5 | 2.5 | Md      | 28km SE of
El Negro, Puerto Rico            |
0101000020E610000062A1D634EF6850C0E561A1D634DF3140
 pr16282000 | 2016-10-08 04:40:43.2+02 |    6 | 2.6 | Md      | 14km SSE
of Tallaboa, Puerto Rico          |
0101000020E610000044696FF085A950C0FF21FDF675E03140
 pr16282000 | 2016-10-08 04:40:43.2+02 |    6 | 2.6 | Md      | 14km SSE
of Tallaboa, Puerto Rico          |
0101000020E610000044696FF085A950C0FF21FDF675E03140
 pr16282001 | 2016-10-08 11:27:19.6+02 |   27 | 2.9 | Md      | 28km NNW
of Charlotte Amalie, U.S. Virgin Islands |
0101000020E6100000DAACFA5C6D4150C0857CD0B359953240
```

> The data is in the WGS84 latitude-longitude coordinate system, so geometry must be cast to geography to get accurate results, as explained in the *Spatial measurement* section.

This is fast for a table with 50 rows, but what if we'd like to reverse geocode (find the nearest address) given a pair of coordinates and an os_address_base_gml table?

```
SELECT * FROM data_import.os_address_base_gml
ORDER BY ST_Distance(wkb_geometry::geography,
ST_SetSRID(ST_MakePoint(-3.5504,50.7220),4258)::geography)
LIMIT 1;
```

This will require a lot of geodesic calculations, as the ST_Distance function cannot use a spatial index.

To make things better, a <-> operator can be used to make indexed K-nearest feature searches. First, let's change the data type of the os_address_base_gml geometry column to geography, as the index won't work when using a type cast:

```
ALTER TABLE data_import.os_address_base_gml ALTER COLUMN wkb_geometry SET
DATA TYPE geography;
```

Then, the query for reverse geocoding will look like this:

```
SELECT * FROM data_import.os_address_base_gml
ORDER BY wkb_geometry <->
ST_SetSRID(ST_MakePoint(-3.5504,50.7220),4258)::geography
LIMIT 1;
```

Using the right data type, a spatial index, and a <-> operator can make this kind of query an order of magnitude faster.

Beware, though--this operator compares the distances between the centers of bounding boxes. For points, this is not a problem, as the BBOX center of the point is equal to that point. But for more complicated shapes the results might be suboptimal. To address this, one can use the <-> operator as a prefilter, and then do a precise computation with ST_Distance:

```
WITH prefilter AS (
  SELECT *, ST_Distance(way, ST_SetSRID('POINT(-100
6705148)'::geometry,900913)) AS dist FROM planet_osm_polygon
  ORDER BY way <-> ST_SetSRID('POINT(-100 6705148)'::geometry,900913) LIMIT
10
)
SELECT * FROM prefilter ORDER BY dist LIMIT 1;
```

Summary

Spatial analysis functions are a key feature for spatial databases, and PostGIS has a very rich set of them. One can compose geometries from raw coordinates, compose geometries into more complex shapes, break down complex shapes into elementary geometries, calculate spatial metrics, and query features based on their location. Spatial analysis functions require the geometries to conform to a specification, so validation and automated repair functions are also provided.

Most spatial analysis functions are easy to understand, but special attention must be paid to coordinate systems. When using latitude-longitude coordinate systems, it's best to use the geography data type to ensure the calculated measurements are correct.

3
Data Processing - Vector Ops

Modern databases shine not only at data storage and retrieval, but also data processing. The set of statistical and mathematical functions in pure PostgreSQL is impressive. In PostGIS, the same principle applies to spatial data. A wide array of geoprocessing functions is at the database user's disposal, and all these functions can be called from within SQL statements. In this chapter, we will discuss the processing capabilities of PostGIS referring as regards to common GIS operations. Those operations include the following:

- Merging and splitting geometries
- Buffering and offsetting geometries
- Computing polygons' centroids and point-on-surfaces
- Computing the difference of geometries
- Reprojecting geometries
- Querying features based on their spatial relationships

Primer - obtaining and importing OpenStreetMap data

In this chapter, the examples we will look at will use spatial data from the OpenStreetMap project, *a Wikipedia for maps*. It's a convenient data source because it's free for everyone to use and download, and it has global coverage (although the quality may vary). The OSM project itself only serves a full dump (called `planet.osm`), which is huge and hard to process. Luckily, third-party services offering smaller extracts exist.

At the time of writing, there were three notable services:

- **Mapzen** (`https://mapzen.com/data/metro-extracts/`) offers ready-made extracts of metropolitan areas around the world for anyone, and custom extracts for registered users.
- **Geofabrik** (`http://download.geofabrik.de/`) offers continental and country extracts. No sign-in is required.
- **BBBike** (`http://extract.bbbike.org/`) offers custom extracts of medium-sized areas (up to 24 million square kilometers or 768 MB of data). No sign-in is required, but a valid e-mail address is. As the extracts are generated on-demand, it takes a couple of minutes to generate them and give a unique URL.

For the purpose of this chapter's examples, let's pick a city or county-sized extract (so BBBike and Mapzen services are a best fit) _in _PBF _format_. This is an OSM-specific exchange format.

After downloading, the file can be imported using the `ogr2ogr` command-line tool:

```
ogr2ogr -t_srs EPSG:32633 -f PostgreSQL "PG:dbname=mastering_postgis
host=localhost user=osm password=osm" planet_17.894_49.888_ef55391f.osm.pbf
```

Replace the database credentials and PBF file name with yours, and the EPSG code to the appropriate projection for your area of interest.

This is convenient, as ogr2ogr is widely used for spatial data conversion, but is not particularly efficient, both in terms of processing power and the disk space required. For larger, country, or continental extracts, or even a full dump, another OSM import tool such as `osm2pgsql` or `Imposm` is required. This is, however, outside of the scope of this book.

The import tool creates one table per geometry type: points, lines, MultiPolygons, and MultiLineStrings, and the columns refer to the most commonly used OSM tags.

Merging geometries

In the previous chapter, we learned how to use `ST_Collect` function to compose Multi-geometries from components. This is computationally cheap, but sometimes, retaining the borders between components (for example, land parcels) is not desirable. This is where `union` functions come into play.

PostGIS has three unioning functions:

- `ST_Union`
- `ST_MemUnion`, which is memory optimized (that is, it will take more time but less memory)
- `ST_UnaryUnion`, which operates at geometry component-level (and hence is more suitable for Multi-geometries)

Merging polygons

The usage of unioning functions is similar to other `spatial aggregate` functions. The first possibility is to supply two geometries. For example, let's pick two town boundaries and simulate the administrative boundary if they were merged:

```
SELECT ST_Union(
    (SELECT wkb_geometry FROM multipolygons WHERE osm_id = '2828737'),
    (SELECT wkb_geometry FROM multipolygons WHERE osm_id = '2828740')
);
```

The resulting polygon has no boundary between neighboring towns:

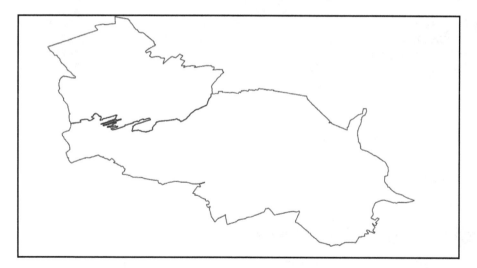

Two polygons about to be merged

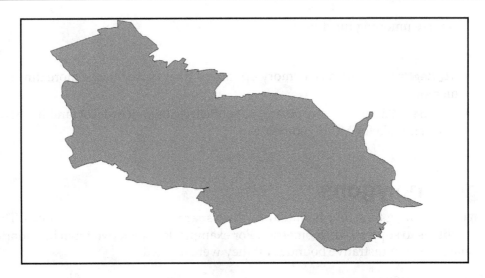

The result of merging

Now we will add two more towns to the stack, using the second possible syntax: providing an array of geometries as an argument:

```
SELECT ST_Union(
ARRAY[
    (SELECT wkb_geometry FROM multipolygons WHERE osm_id = '2828737'),
     (SELECT wkb_geometry FROM multipolygons WHERE osm_id = '2828740'),
   (SELECT wkb_geometry FROM multipolygons WHERE osm_id = '2828739'),
    (SELECT wkb_geometry FROM multipolygons WHERE osm_id = '2828738')

]
);
```

Finally, let's try to merge all boundaries into one big polygon:

```
SELECT ST_Union(wkb_geometry) FROM multipolygons WHERE
boundary='administrative';
```

Merging MultiLineStrings

PostGIS can also merge MultiLineStrings into single-part LineStrings. This is particularly useful when the geometry is to be further used in another function which doesn't accept Multi-geometries as arguments. For this to work, however, the input MultiLineString must not have gaps.

ogr2ogr converts OSM relations (road, rail, tourist routes) into MultiLineStrings. However, not every relation can be *sewed together* using ST_LineMerge. Single-track railways or single-carriageway roads are usually a good example, but some relations, such as dual carriageways, consist of a collection of disjointed lines. In that case, ST_LineMerge will just do nothing and return the original MultiLineString.

In this example, a train route composed of multiple sections of track will be *sewed together* to form a single-part LineString.

The textual representation of an original MultiLineString is as follows:

```
SELECT ST_AsText(wkb_geometry) FROM multilinestrings WHERE osm_id =
'4581657';

st_astext
MULTILINESTRING((18.2172473 50.0790981,18.2169285 50.0788042,18.2166262
50.0785414,18.2161234 50.0780898,18.2154094 50.0774614,18.2150988
50.0772028,18.2148354 50.0770007,18.2145185 50.0767715,18.2141338
50.0765137,18.2137063 50.0762518,18.2131764 50.0759516))
```

After merging, it looks like this:

```
SELECT ST_AsText(ST_LineMerge((SELECT wkb_geometry FROM multilinestrings
WHERE osm_id = '4581657')));

st_astext
LINESTRING(18.2282796 50.0923211,18.2280783 50.0922193,18.2278694
50.0921051,18.2276675 50.0919549,18.2274702 50.0917769,18.227307
50.0916126,18.2271524 50.0914074,18.2260282 50.0899124,18.225715
50.0895162,18.2255041 50.0892647,18.2253105 50.0890423,18.2250587
50.0887412,18.2245293 50.0880549,18.2237535)
```

The MultiLineString doesn't have to exist in a database in ready-made form - it can also be created on the fly using ST_Collect:

```
SELECT ST_LineMerge(ST_Collect(wkb_geometry)) FROM lines WHERE
waterway='river' AND name = 'Odra';
```

This will merge multiple sections of a single river into a single LineString geometry:

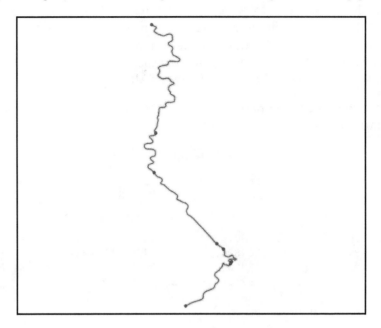

The result of the ST_LineMerge function. The red dots indicate the starting points of consecutive parts of the MultiLineString

Slicing geometries

PostGIS is good at splitting geometries too. There are methods for splitting line and polygon geometries using a second geometry (blade), and for extracting sections of linear features based on distance.

Splitting a polygon by LineString

In our first example, we'll split a polygon (county boundary) using a line (river). We will use an ST_Split function for that.

This function accepts two arguments: a geometry to be split, which can be of the (Multi)Polygon or (Multi)LineString type, and a blade, which can be a LineString for polygons and LineStrings or a Point for lines:.

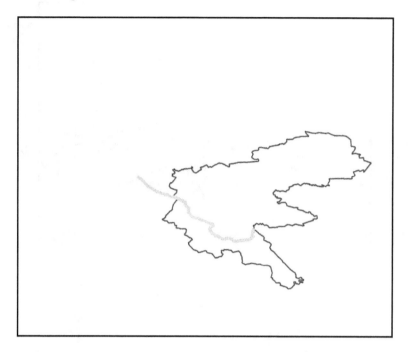

A polygon about to be split with a LineString used as a blade

```
SELECT ST_Split(
  (SELECT wkb_geometry FROM multipolygons WHERE osm_id = '2417246'),
  (SELECT wkb_geometry FROM lines WHERE osm_id = '224461074'))
```

This yields a GeometryCollection, which isn't supported in most GIS software, including QGIS. To extract individual parts after splitting, and to visualize the result in QGIS, we'll need to use the ST_Dump function:.

```
SELECT (ST_Dump(ST_Split(
  (SELECT wkb_geometry FROM multipolygons WHERE osm_id = '2417246'),
  (SELECT wkb_geometry FROM lines WHERE osm_id = '224461074')))).geom
```

This converts a `GeometryCollection` into a set of rows, each with a polygon geometry, which can be visualized by any PostGIS-supporting software without any problems.

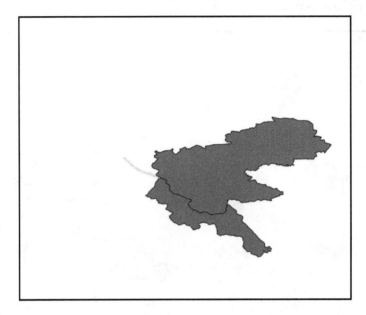

Two polygons after executing an ST_Split function

Splitting a LineString with another LineString

In this example, we'll split a linear feature (a road) with another road at an intersection:

```
SELECT (ST_Dump(ST_Split(
 (SELECT wkb_geometry FROM multilinestrings WHERE osm_id = '333295'),
 (SELECT wkb_geometry FROM multilinestrings WHERE osm_id =
'2344149')))).geom
```

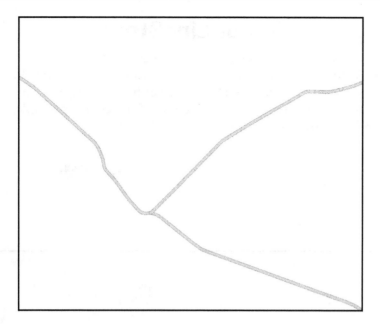

Two intersecting LineStrings

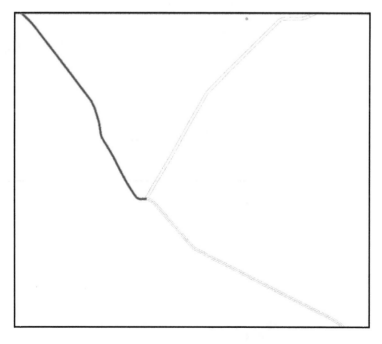

A LineString split using another LineString at an intersection

Extracting a section of LineString

LineStrings can be sliced not only using other geometries, but also based on a fraction of their length. This is done by the ST_LineSubstring function. It accepts three arguments: an input geometry, which must be a LineString (a MultiLineString must be merged using ST_LineMerge), the starting fraction, and the ending fraction. For example, we can extract the first half of the Odra river as follows:

```
SELECT ST_LineSubstring(
    (SELECT ST_LineMerge(ST_Collect(wkb_geometry)) FROM lines WHERE
name='Odra' AND waterway='river'),
    0,
    0.5
)
```

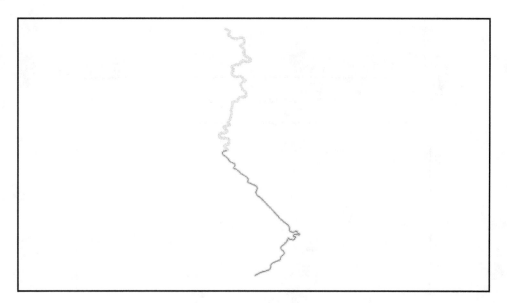

A line cut in half with an ST_LineSubstring

When distance values are needed, the ST_Length function can be used to calculate the necessary fraction. For example, say that railway line No. 177 has been scheduled for repair between the 13th and 24th kilometer. How do we draw this section on a map?

```
WITH line_geom AS (SELECT ST_LineMerge((SELECT wkb_geometry FROM
multilinestrings WHERE osm_id = '4581657')))
SELECT ST_LineSubstring(
  line_geom.st_linemerege,
  13000 / ST_Length(line_geom.st_linemerge),
  24000 / ST_Length(line_geom.st_linemerge)
  ) FROM line_geom;
```

The WITH clause is called a **common table expression** (**CTE**). It allows us to store the geometry in question as a virtual table after merging as a virtual table, so we don't have to type SELECT ST_LineMerge... and its ID over and over again.

The geometry is stored in the EPSG:32633 coordinate system, so its units are meters. Therefore, we need to multiply the kilometer values by 1,000. The result looks like the following:

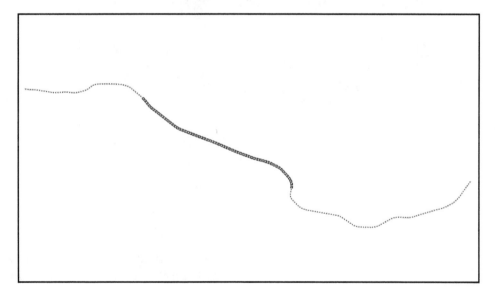

A Lline section extracted by the starting and ending distance

Buffering and offsetting geometries

A buffer is a very common GIS operation. PostGIS can create polygonal buffers from any geometry with configurable distance and approximation levels.

For example, a simple 1,000 meter buffer from a Point looks like the following:

```
SELECT ST_Buffer(
  (SELECT wkb_geometry FROM points WHERE osm_id = '253525668'),
  1000);
```

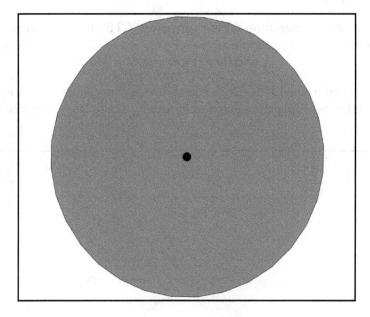

A simple buffer with default parameters

The first argument is an input geometry, and the second is the buffer distance in the units of the geometry's coordinate system.

 If the geometry is in a latitude-longitude coordinate system, then cast the geometry to a Geography type in order to be able to give the distance in meters.

The default buffer uses eight segments to approximate a quarter circle. If it's too coarse, more segments can be introduced at the expense of processing power:

```
SELECT ST_Buffer(
  (SELECT wkb_geometry FROM points WHERE osm_id = '253525668'),
  1000,32);
```

We can also do the opposite, such as when we create an octagonal buffer with only 2 segments per quarter circle:

```
SELECT ST_Buffer(
  (SELECT wkb_geometry FROM points WHERE osm_id = '253525668'),
  1000,2);
```

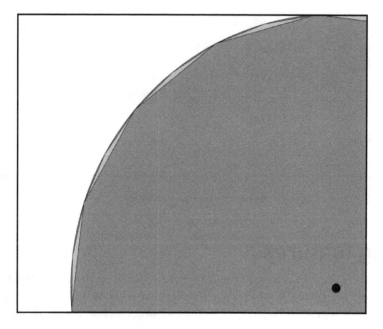

The boundaries of an ST_Buffer output dependent on segment count. Green -- 4 segments., yellow -- 8 segments,. red -- 32 segments

 While buffers can be used for doing radius queries (find all restaurants within 1,000 meters of my location), PostGIS has a specialized function for that, which is about an order of magnitude faster. Refer to the ST_DWithin function description in the *Spatial relationships* section.

To replicate the **dissolve result** option found in Desktop GIS software, ST_Buffer can be used in conjunction with ST_Union. For example, to create a 200 meter buffer zone from all rivers as one big MultiPolygon, enter the following:

```
SELECT ST_Union(
    ST_Buffer(wkb_geometry,200)
) FROM lines WHERE waterway = 'river';
```

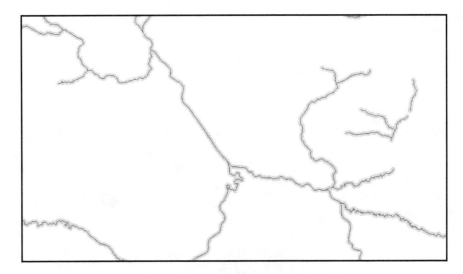

The result of using ST_Unioning multiple buffers

Offsetting features

The offset function is used to create a linear feature parallel to the original line at a specified offset. For example, let's imagine that we are designing a second track for a railway line. As a base, we will create a geometry parallel to the existing track at a 4 meter offset:

```
SELECT ST_OffsetCurve(ST_LineMerge(wkb_geometry),4) FROM multilinestrings
WHERE osm_id = '4581657';
```

Despite its name, the ST_OffsetCurve function operates on ordinary LineStrings. The MultiLineStrings must be merged with ST_LineMerged prior to analysis:

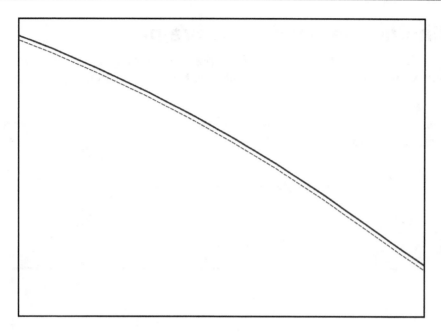

An original line (solid) and offset (dashed)

The `ST_OffsetCurve` function creates a new geometry on the left side of the original geometry, maintaining its direction. To create a new geometry to the right, simply supply a negative offset value.

This function also offers further fine-tuning options, including the following:

- `quad_segs`: Number of segments to approximate a quarter circle (just like the third parameter for `ST_Buffer`), defaults to 8
- `join`: One of round, `mitre`, bevel - line join style
- `mitre_limit`: Mitre ratio limit for `mitre` join

The fine-tuning parameters are given as a third argument, in the form of a space-separated string. So, a 32-segment quarter circle has a `mitre` join and a `mitre` limit of 2:

```
SELECT ST_OffsetCurve(ST_LineMerge(wkb_geometry),4,'quad_segs=32 join=mitre
mitre_limit=2') FROM multilinestrings WHERE osm_id = '4581657';
```

Creating convex and concave hulls

Another possibility for creating polygons from points is to compute a convex or concave hull. A convex hull is a polygon containing all input points. It's often described as a *rubber band*.

PostGIS has an ST_ConvexHull function. It takes one argument - an input geometry. It's not an aggregate function, so in order to supply a set of points, the ST_Collect aggregate has to be used first.

The convex hull of all mineshafts is written as follows:

```
SELECT ST_ConvexHull(
    ST_Collect(wkb_geometry)
) FROM points WHERE man_made = 'mineshaft'
```

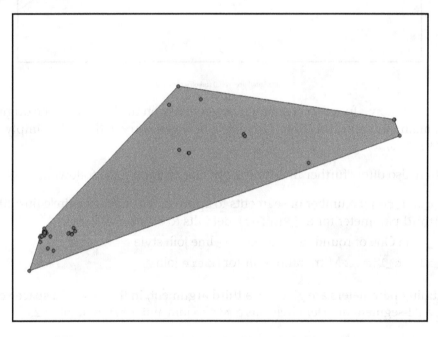

A set of points and a convex hull

The convex hull is cheaply computed, but it's rarely a good approximation. Distant and isolated features extend the hull, covering areas where the phenomenon isn't present. To address this problem, another algorithm, called a concave hull, was created.

In PostGIS, the implementation is provided by the ST_ConcaveHull function. Apart from the geometry, it takes two additional parameters: the concavity factor (between 0 and 1) controlling the maximal fraction of the convex hull's area that the concave hull can have, and a Boolean (TRUE or FALSE, default FALSE) controlling whether the output polygon can have holes:

```
SELECT ST_ConcaveHull(
    ST_Collect(wkb_geometry),
    0.5,
    TRUE
) FROM points WHERE man_made = 'mineshaft'
```

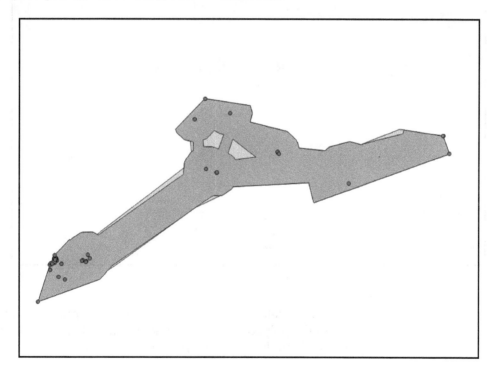

A set of points and concave hulls with (darker) and without (lighter) holes

Computing centroids, points-on-surface, and points-on-line

Whether it's needed for dataset generalization, labelling, or using in a point-only calculation, the centroid calculation right in the DB might come in handy. This is the job of the ST_Centroid function. The usage is simple, as the only required parameter is the input geometry, which can be of any type (an ST_Centroid of a point will return the original point). However, the most common (and legitimate) use case for centroid calculation is for polygons. Let's compute centroids for all water bodies in our database:

```
SELECT ogc_fid, name, ST_Centroid(wkb_geometry) FROM multipolygons WHERE
"natural" = 'water';
```

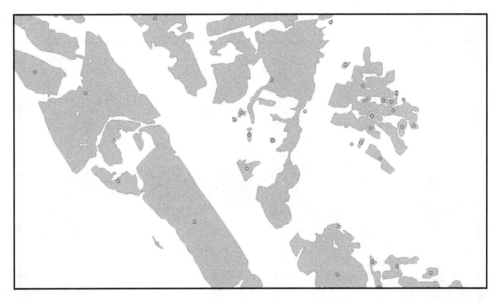

Polygons and their centroids

But for some features, such as an oxbow lake, the centroid isn't a correct representation because it can be located outside of the polygon's interior. Also, for lines, the centroid can be computed, but the resultant point will rarely be located precisely on a line.

For these cases, another function, ST_PointOnSurface, was designed. The point-on-surface is guaranteed to be contained within the polygon's interior (or along a line, in case a LineString is supplied):

```
SELECT ogc_fid, name, ST_PointOnSurface(wkb_geometry) FROM multipolygons
WHERE "natural" = 'water';
```

Centroids (lighter/yellow) and points-on-surface (darker/green)

Reprojecting geometries

Reprojecting, also called transformation, is a process of converting a geometry's coordinates from one coordinate system to another. Since PostGIS spatial analysis functions can't operate on geometries with different coordinate systems, it's a very important functionality. It's provided by the ST_Transform function.

ST_Transform accepts two arguments: the input geometry and the target SRID. For example, to transform the coordinates of a point feature (city center) to latitude-longitude, we write the following:

```
SELECT ST_AsText(ST_Transform(wkb_geometry,4326)) FROM points WHERE osm_id
= '253525668';

st_astext
POINT(18.5419933 50.0955793)
```

While `ST_Transform` can be used on the fly, it's computationally expensive and can cause a complex query to run very slowly. If it's necessary to run a spatial analysis using tables with different SRIDs, it's will be wise to create a materialized view with geometries reprojected beforehand:

```
CREATE MATERIALIZED VIEW boundaries_3857 AS SELECT ogc_fid, name, boundary,
admin_level, ST_Transform(wkb_geometry,3857) AS wkb_geometry FROM
multipolygons WHERE boundary='administrative';
```

This will create a materialized view containing a subset of the `multipolygons` table (administrative boundaries) with their geometries reprojected into the *web mercator* coordinate system, which has an SRID of `3857`.

For efficient querying, this materialized view should also have a spatial index:

```
CREATE INDEX boundaries_3857_sidx ON boundaries_3857 USING
GIST(wkb_geometry);
```

Now, any spatial queries with features in the EPSG:3857 coordinate system will run as smoothly as possible.

Geometries can also be reprojected in place, for example:

```
UPDATE multilinestrings SET wkb_geometry = ST_Transform(wkb_geometry,3857);
```

But this will fail if a geometry column has a typmod indicating the only possible SRID, which is true in our case, and the query will result in an error:

```
Geometry SRID (3857) does not match column SRID (32633)
```

To work around this, a typmod can be removed with an `ALTER COLUMN` statement:

```
ALTER TABLE multilinestrings ALTER COLUMN wkb_geometry SET DATA TYPE
geometry;
```

Beware, though - this will remove the protection from accidentally inserting a geometry in another SRID into the table.

Spatial relationships

In the previous chapter, one spatial relationship function - the `ST_Intersects`, was introduced. This is only the beginning of the story; there are many more, finer functions available, to explore the relationships between features more precisely. For more details, read on.

Touching

A geometry is defined as touching when at least its vertex or edge lies on the edge of the reference geometry, but no points are inside its interior.

For example, let's find all the tributaries of the Odra river:

```
SELECT name,waterway FROM lines WHERE waterway in ('river','stream') AND
name != 'Odra' AND ST_Touches(
   wkb_geometry,
   (SELECT ST_Collect(wkb_geometry) FROM lines WHERE name = 'Odra')
   );
```

The `ST_Collect` is needed for collecting all sections into a single geometry (`ST_Touches` accepts MultiLineStrings, so there's no need for Union or LineMerge there), and the `name != 'Odra'` condition is needed to exclude neighbouring sections of the same river.

Crossing

A **geometry** is defined as crossing when at least one vertex of it lies in the interior of the reference geometry (when it's a polygon), or crosses the LineString. An example of a query using this relationship is finding all the bridges crossing the Odra river:

```
SELECT name,wkb_geometry FROM lines WHERE other_tags LIKE '%bridge%' AND
ST_Crosses(wkb_geometry,
   (SELECT ST_Collect(wkb_geometry) FROM lines WHERE waterway='river' AND
   name='Odra'));
```

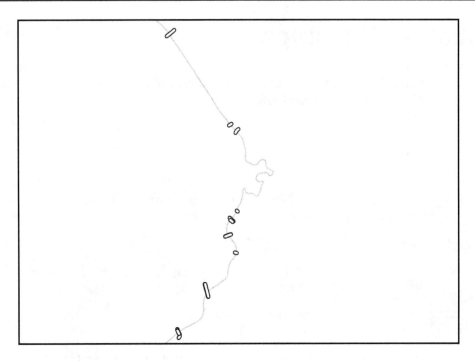

A set of LineStrings fulfilling the ST_Crosses condition

Overlapping

Overlapping geometries share some portion of their interiors, but one can't completely cover another. This relationship is tested with the ST_Overlaps function. This query will reveal all bus routes that share some part with the E-3 route:

```
SELECT ogc_fid, other_tags, wkb_geometry FROM multilinestrings WHERE
ST_Overlaps(
    wkb_geometry,
    (SELECT wkb_geometry FROM multilinestrings WHERE other_tags LIKE
'%E-3%')
);
```

Please note that, if a route exactly duplicates the path, or is a subsection that is completely within the bigger route, it will not be included in the result:

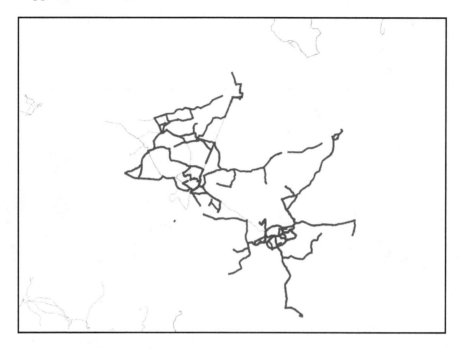

A bus route network: the E-3 route (red), overlapping routes (violet), and other routes (gray)

Containing

In contrast to crossing and overlapping, containing means a geometry must be completely covered by the reference geometry, and no point may be located outside. To test this, two functions are provided: ST_Within and ST_Contains. The only difference is the order of arguments.

First, we will use the `ST_Within` function to determine how many water bodies are within the boundaries of Rybnik (which literally means *fish pond* in Czech, by the way):

```
SELECT name,wkb_geometry FROM multipolygons WHERE "natural"='water' AND
ST_Within(wkb_geometry,
(SELECT wkb_geometry FROM multipolygons WHERE name='Rybnik' AND
admin_level='7'));
```

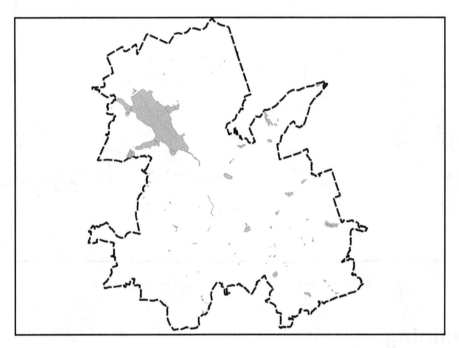

The result of an ST_Within query - a set of polygons fully contained within another polygon

Then we will deal with the second query - finding power towers in `Rudnik` using `ST_Contains`:

```
SELECT name,wkb_geometry FROM points WHERE other_tags LIKE '%power%' AND
ST_Contains(
(SELECT wkb_geometry FROM multipolygons WHERE name='gmina Rudnik' AND
admin_level='7'),wkb_geometry);
```

Did you notice the difference? When using `ST_Within`, the geometries to be found are supplied as the first argument, and the boundary to search within as the second. For `ST_Contains`, the search boundary was first, and the geometries to be found, second. If we swapped the arguments, the query would return 0 rows: no polygon (at least, not a valid one) will fit into a point.

Radius queries

For finding features located within a specified distance, PostGIS is equipped with a specialized ST_DWithin function. It can be imagined as a combination of ST_Buffer and ST_Within, only it omits the buffer geometry construction step and is much faster.

The syntax is as follows:

```
ST_DWithin(tested_geometry,reference_geometry,distance)
```

Let's look at an example - listing all villages located within 10 kilometers of Rybnik city:

```
SELECT name,wkb_geometry FROM points WHERE place='village' AND ST_DWithin(
    wkb_geometry,
    (SELECT wkb_geometry FROM points WHERE name='Rybnik' AND place='city'),
    10000
)
```

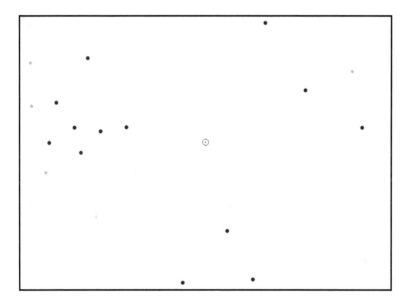

Rybnik city (in the middle), villages within 10 kilometers (black), and other villages (gray)

Summary

Vector operation functions can be used as a rich geoprocessing toolbox, all without leaving the database environment. Using nothing but Spatial SQL, one can split and merge features, reproject them, create buffers and hulls, offset geometries, and much more. And when it comes to querying, PostGIS offers fine-grained spatial relationship functions to pick only specific features. This wouldn't be possible without a rich open source GIS ecosystem; the geoprocessing and spatial relationship functions are available thanks to the GEOS library, and reprojections are done with the help of the PROJ.4 library. Other database vendors often choose to implement spatial functions from scratch, and stick to a moderate set of functions defined in the OGC specification. The geoprocessing toolbox is one of PostGIS's greatest strengths. This toolbox is not limited to vector data manipulation. In the next chapter, we will discuss the functions available for raster data processing.

4
Data Processing - Raster Ops

PostGIS raster's goal is to implement the raster type as much as possible like the geometry type is implemented in PostGIS and to offer a single set of overlay SQL functions (such as `ST_Intersects`) operating seamlessly on vector and raster coverages.

Each raster or raster tile is stored as a row of data in a PostgreSQL database table. It is a complex type, embedding information about the raster itself (width, height, number of bands, pixel type for each band, and no data value for each band) along with its geolocalization (pixel size, upper left pixel center, rotation, and SRID). These metadata are accessible by `raster_columns` view.

Something that shows flexibility of PostGIS raster support is in-db / out-db raster tile storage. Operations on these are identical, no matter if the raster is stored internally in PostgreSQL or file in filesystem. The only drawback is query processing time for out-db raster coverages.

PostGIS Raster is expressed in different forms, depending on the level at which it is referred:

- **WKT**: **Well Known Text** refers to the human readable text format used when inserting a raster with `ST_RasterFromText()` and retrieving a raster with `ST_AsText()` (these two are not implemented yet). This format can result in the loss of precision when used with floating point values. This is why the HEXWKB form is preferred when importing/exporting in textual form.
- **WKB**: **Well Known Binary** refers to the binary equivalent of the WKT. It is used when inserting a raster using `ST_RasterFromWKB()` and retrieving a raster using `ST_AsBinary()`.

- **HEXWKB: Hexadecimal WKB** is an exact hexadecimal representation of the WKB form. It is also called the canonical form. It is what you get from the loader (`raster2pgsql`), what is accepted by the raster type input function (`ST_Raster_In`), and what you get when outputting the value of a raster field without conversion (for example, `SELECT rast FROM table`).
- **Serialized**: The serialized format is what is written to the filesystem by the database. It differs from WKB in that it does not have to store the endianness of the data and that it must be aligned. Serializing is the action of writing data to the database file, deserializing is the action of reading this data.

As a convention similar to geometry columns (column named `geom`), the `rast` column is used for storing raster datatype. The result of raster conversion functions maybe returned as the function name; it's our role to set it with AS keyword.

There are a variety of PostGIS raster functions especially created for raster analysis and processing. We can divide them into some groups:

- Raster table management
- Raster constructors
- Dataset accessors and editors
- Band accessors and editors
- Pixel accessors and editors
- Raster geometry editors
- Outputs
- Processing tools

Preparing data

Raster import was covered in `Chapter 1`, *Importing Spatial Data*, subject and exporting is on its way in the next chapter, so we can look strictly into vector <-> raster conversion and processing.

For our use case, we will use EU-DEM dataset, produced using Copernicus data and information funded by European Union - EU-DEM layers. It's one arc second model of the whole European Union (without overseas territories). With its open license, we can use it with no concerns. The dataset is accessible by `http://www.eea.europa.eu/data-and-maps /data/eu-dem`.

As a vector layer, we can use OpenStretMap data for the Poland and Czech Republic administrative units.

In the first step, let's import the dataset as we learned in `Chapter 1`, *Importing Spatial Data*. For optimization of import processes, look into metadata, for SRID and size of raster, `gdalinfo` makes it easy:

```
user@machine:~/dev/gis/dem$ gdalinfo eudem_dem_5deg_n45e015.tif
Driver: GTiff/GeoTIFF
Files: eudem_dem_5deg_n45e015.tif
Size is 18000, 18000
Coordinate System is:
GEOGCS["ETRS89",
    DATUM["European_Terrestrial_Reference_System_1989",
        SPHEROID["GRS 1980",6378137,298.2572221010042,
            AUTHORITY["EPSG","7019"]],
        AUTHORITY["EPSG","6258"]],
    PRIMEM["Greenwich",0],
    UNIT["degree",0.0174532925199433],
    AUTHORITY["EPSG","4258"]]
Origin = (15.000000000000000,50.000000000000000)
Pixel Size = (0.000277777777778,-0.000277777777778)
Metadata:
  AREA_OR_POINT=Area
Image Structure Metadata:
  COMPRESSION=LZW
  INTERLEAVE=BAND
Corner Coordinates:
Upper Left  (  15.0000000,  50.0000000) ( 15d 0' 0.00"E, 50d 0' 0.00"N)
Lower Left  (  15.0000000,  45.0000000) ( 15d 0' 0.00"E, 45d 0' 0.00"N)
Upper Right (  20.0000000,  50.0000000) ( 20d 0' 0.00"E, 50d 0' 0.00"N)
Lower Right (  20.0000000,  45.0000000) ( 20d 0' 0.00"E, 45d 0' 0.00"N)
Center      (  17.5000000,  47.5000000) ( 17d30' 0.00"E, 47d30' 0.00"N)
Band 1 Block=18000x1 Type=Float32, ColorInterp=Gray
  Min=54.810 Max=2424.210
  Minimum=54.810, Maximum=2424.210, Mean=317.651, StdDev=254.987
  NoData Value=nan
  Metadata:
    STATISTICS_MAXIMUM=2424.2099609375
    STATISTICS_MEAN=317.65072388322
    STATISTICS_MINIMUM=54.810001373291
    STATISTICS_STDDEV=254.98702538825
```

Now we know that EPSG:4258 (ETRS89) should be taken and tiling applied; in our case, 4500 px will be good approach. Using copy, it will be done faster.

```
raster2pgsql -s 4258 -C -I -l 2,4,8 -t 4500x4500 -F -Y *.tif raster_ops.dem
| psql -d mastering_postgis -U osm
```

 We need to provide port and host name to psql for the command to work.

If you don't use -C flag during import, probably the AddRasterConstraints() function will be used later to fill in correct values in the raster_columns catalog.

Processing and analysis

Now we are ready to start processing our dataset and make some scientific analysis.

The most common use cases of raster data are imagery, DEM, and statistical data. Right now, we'll look closer to Digital Elevation Model processing.

We can use the following:

- ST_Slope
- ST_Hillshade
- ST_Aspect
- ST_TPI
- ST_TRI
- ST_Roughness

All these functions are internally realized as MapAlgebra Callback functions. Later we will see how to do it on our way. So, calculate the slope values of our DEM data. The function syntax is like this:

```
ST_Slope(raster rast, integer nband=1, text pixeltype=32BF, text
units=DEGREES, double precision scale=1.0, boolean
interpolate_nodata=FALSE);
```

We should especially pay attention to scale parameter. In case of geographical coordinate reference systems (such as our imported data), when units of distance are in degrees, and elevation in meters, use scale=111120, or in case of imperial units set it to 370400. Units parameter offers DEGREE, RADIANS, or PERCENT as a result of calculations. Because there are two variants of this function, arguments must be explicitly cast.

For calculation time reduction, I will use the clipped version of the dem table named clip. We will return to examples of extraction raster parts later in this chapter:

```
SELECT rid, ST_Slope(rast::raster,
'1'::int,'32BF'::text,'DEGREES'::text,'111120'::double precision) AS rast
FROM raster_ops.clip WHERE rid='4';
```

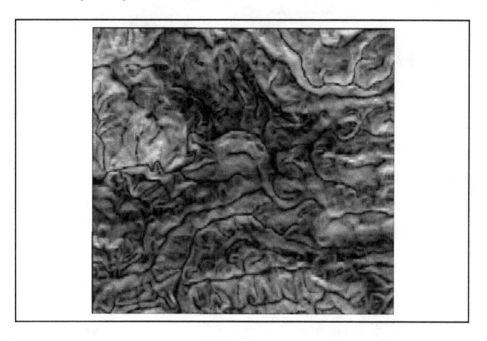

By analogue to slope calculations, aspect function is executed as follows, and the output raster will have aspect values in each pixel, in units specified in the fourth argument:

```
SELECT rid, ST_Aspect(rast::raster,
'1'::int,'32BF'::text,'DEGREES'::text,'111120'::double precision) FROM
raster_ops.clip WHERE rid='4';
```

More arguments must be set on ST_Hillshade functions.

```
ST_HillShade(raster rast, integer band=1, text pixeltype=32BF, double
precision azimuth=315, double precision altitude=45, double precision
max_bright=255, double precision scale=1.0, boolean
interpolate_nodata=FALSE);
```

We can execute a query with only raster column and band and it will work. But for LatLon datasets again, every argument must be explicitly set and cast to type.

Sometimes we need to change values of our dataset in an organized way, not in algebraic way, but as some groups-classes. There's a reclassify tool in PostGIS for that:

```
SELECT rid,ST_Reclass(
(ST_SLOPE(rast::raster,'1'::int,'32BF'::text,'DEGREES'::text,'111120'::doub
le precision))::raster,
   '1'::int,    '0-5):1, 5-10:2, [10-15):3, 15-25):4,
[25-90):5'::text,'32BF'::text)
FROM eudem.clip;
```

Pay attention to class borders and brackets. In this example, we are reclassifying the ST_slope function results to five classes of slope:

```
0-4.99
5-9.99
10-14.99
15-24.99
25-90 degrees
with 1-5 values.
```

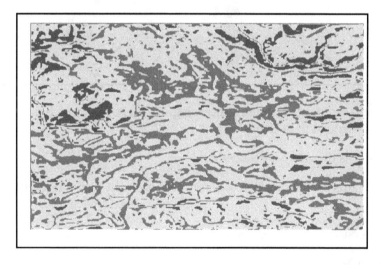

For ST_TRI() and ST_TPI(), the case is more easier as the only argument for these functions is raster column.

TPI measures the relative topographic position of the central point as the difference between the elevation at this point and the mean elevation within a predetermined neighbourhood. Using TPI, landscapes can be classified in slope position classes. TPI is only one of a vast array of morphometric properties based on neighbouring areas that can be useful in topographic and DEM analysis (see Gallant and Wilson, 2000) *Gallant, J.C., Wilson, J.P., 2000. Primary topographic attributes. In: Wilson, J.P., Gallant, J.C. (Eds.), Terrain Analysis: Principles and Applications. Wiley, New York, pp. 51-85.*

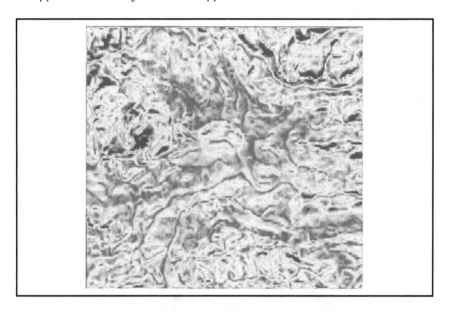

Now, let's look into more complicated processing. As there's not much defined functions for DEM, sometimes we need to define it.

`ST_MapAlgebra()` is the tool that is needed.

There are two variants of this function `expression` and `callbacfunc` with one or two rasters.

The Expression version syntax is as follows:

```
raster ST_MapAlgebra(raster rast, integer nband, text pixeltype, text
expression, double precision nodataval=NULL);
raster ST_MapAlgebra(raster rast1, integer nband1, raster rast2, integer
nband2, text expression, text pixeltype=NULL, text extenttype=INTERSECTION,
text nodata1expr=NULL, text nodata2expr=NULL, double precision
nodatanodataval=NULL);
```

`extenttype` values allowed:

- **INTERSECTION**: The extent of the new raster is the intersection of the two rasters. This is the default.
- **UNION**: The extent of the new raster is the union of the two rasters.
- **FIRST**: The extent of the new raster is the same as the one of the first raster.
- **SECOND**: The extent of the new raster is the same as the one of the second raster.

These values are permitted for expression:

- `[rast1]` or `[rast1.val]`: This is the pixel value of the pixel of interest from `rast1`
- `[rast1.x]`: This is the 1-based pixel column of the pixel of interest from `rast1`
- `[rast1.y]`: This is the 1-based pixel row of the pixel of interest from `rast1`
- `[rast2]` or `[rast2.val]`:This is the pixel value of the pixel of interest from `rast2`
- `[rast2.x]`: This is the 1-based pixel column of the pixel of interest from `rast2`
- `[rast2.y]`: This is the 1-based pixel row of the pixel of interest from `rast2`

Our first expression is as follows:

```
CREATE TABLE ceiled AS SELECT ST_MapAlgebra(rast, 1, NULL,
'ceil([rast]*[rast.x]/[rast.y]+[rast.val])') FROM eudem.clip
```

The result rendered by QGIS is as follows:

For `callbackfunc` version use trickier:

```
raster ST_MapAlgebra(raster rast1, integer nband1, raster rast2, integer
nband2, regprocedure callbackfunc, text pixeltype=NULL, text
extenttype=INTERSECTION, raster customextent=NULL, integer distancex=0,
integer distancey=0, text[] VARIADIC userargs=NULL);
```

Built-in `callbackfunc`:

- `ST_Distinct4ma`: This is the raster processing function that calculates the number of unique pixel values in a neighborhood
- `ST_InvDistWeight4ma`: This is the raster processing function that interpolates a pixel's value from the pixel's neighborhood
- `ST_Max4ma`: This is the raster processing function that calculates the maximum pixel value in a neighborhood
- `ST_Mean4ma`: This is the raster processing function that calculates the mean pixel value in a neighborhood
- `ST_Min4ma`: This is the raster processing function that calculates the minimum pixel value in a neighborhood
- `ST_MinDist4ma`: This is the raster processing function that returns the minimum distance (in number of pixels) between the pixel of interest and a neighboring pixel with value
- `ST_Range4ma`: This is the raster processing function that calculates the range of pixel values in a neighborhood
- `ST_StdDev4ma`: This is the raster processing function that calculates the standard deviation of pixel values in a neighborhood
- `ST_Sum4ma`: This is the raster processing function that calculates the sum of all pixel values in a neighborhood

As an example, use `st_mean4ma`:

```
SELECT     rid,     st_mapalgebra(rast, 1, 'st_mean4ma(double
precision[][][],integer[][],text[])'::regprocedure, '32BF'::text,
'FIRST'::text, NULL) FROM raster_ops.clip
```

We are not bound to these built in functions-we can prepare our own PL/pgSQL callback-just remember that it should have input arguments and should return the following:

```
'sample_callbackfunc(double precision[], integer[], text[])'::regprocedure
```

Analytic and statistical functions

Do you remember ST_Reclass? How about getting good classes? There are two very interesting tools in PostGIS. ST_Histogram() and ST_Quantile(). First, return SETOF record of the defined number of histogram bins. The only thing we need to set is the raster column, band, and number of bins. We can break results into min, max, count, and percent with SELECT (histogram).* syntax.

```
mastering_postgis=# SELECT (stat).* FROM (SELECT ST_Histogram(rast,1,6) AS
stat FROM eudem.clip) AS foo;
      min          |        max          | count |        percent
-------------------+---------------------+-------+----------------------
 440.980010986328  | 549.598332722982    | 19517 |  0.103899491602119
 549.598332722982  | 658.216654459635    | 76015 |  0.404668742846496
 658.216654459635  | 766.834976196289    | 55312 |   0.29445553514866
 766.834976196289  | 875.453297932943    | 26941 |  0.14342143788762
 875.453297932943  | 984.071619669597    |  8320 |  0.0442918363544412
 984.071619669597  | 1092.68994140625    |  1740 | 0.00926295616066438
(6 rows)
```

The second tool helps us to select class breaks. Here, the execution is even easier-only raster column and band are necessary.

```
mastering_postgis=# SELECT (stat).* FROM (SELECT ST_Quantile(rast,1) AS
stat FROM eudem.clip) AS foo;
 quantile |      value
----------+------------------
        0 | 440.980010986328
     0.25 | 593.859985351562
      0.5 | 655.929992675781
     0.75 | 742.580017089844
        1 | 1092.68994140625
(5 rows)
```

It is possible to set real quantile percents and get values with ARRAY[].

```
mastering_postgis=# SELECT (stat).* FROM (SELECT
ST_Quantile(rast,1,ARRAY[0.3,0.6,0.88]) AS stat FROM eudem.clip) AS foo;
 quantile |      value
----------+------------------
      0.3 | 606.700012207031
      0.6 | 684.010009765625
     0.88 | 806.630004882812
(3 rows)
```

Vector to raster conversion

The most common method to change vector geometries to raster is using the ST_AsRaster function. The only drawback is that it needs reference raster geometry.

```
CREATE TABLE raster_ops.admin_rast AS
SELECT ST_Union(ST_AsRaster(geometry, rast, '32BF', '1', -9999)) rast
FROM raster_ops.admin, (SELECT rast FROM eudem.clip LIMIT 1) rast;
```

This function allows you to set values of raster pixels according to attribute value, in our example, the height column.

```
CREATE TABLE raster_ops.admin_rast AS
SELECT ST_Union(ST_AsRaster(geometry, rast, '32BF', height, -9999)) rast
FROM raster_ops.admin, (SELECT rast FROM eudem.clip LIMIT 1) rast;
```

Raster to vector conversion

The most useful function there for raster values to Vector is ST_DumpAsPolygons(). As a result we get geomval (set-of-records).

```
SELECT rid, (foo).*
   FROM (SELECT rid,
(ST_DumpAsPolygons(ST_Reclass((ST_SLOPE(rast::raster,'1'::int,'32BF'::text,
'DEGREES'::text,'111120'::double precision))::raster, '1'::int,'0-5):1,
```

```
5-15]:2, [10-15:3, [15-25):4, 25-90]:5'::text,'32BF'::text))) FROM
eudem.clip) AS foo;
```

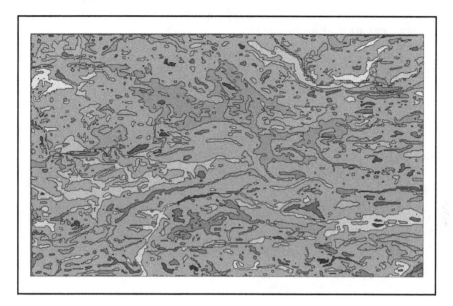

Another way to get some information is to query for values of raster cell at point of interest.
As an example, let's check the value of DEM height for Istebna commune center:

```
mastering_postgis=# SELECT ST_Value(c.rast,(SELECT geom FROM
raster_ops.places WHERE name='Istebna')) FROM eudem.clip c;
     st_value
-------------------
 661.260009765625
(1 row)
```

Spatial relationship

The most common need for extraction of some raster data from larger dataset is to reduce
computing time. Let's look into spatial relationship and extraction functions in PostGIS. For
getting results in raster space, we should use ST_Clip() and ST_Intesection() for
vector space results.

In the previous example, we extracted one raster cell value for vector point. Now, we need to query for multiple point values. There we should use the `ST_Intersection()` function. As a standard, this function returns geoval type in vector space, that we need to join with our vector points. Let's look at an example:

```
SELECT  foo.name,
   (foo.geomval).val AS height
FROM (
  SELECT
    ST_Intersection(A.rast, g.geom) As geomval,
    g.name
  FROM eudem.clip AS A
  CROSS JOIN (
    SELECT geom, name FROM raster_ops.places WHERE type='village'
  ) As g(geom, name)
  WHERE A.rid = 4
) As foo;
    name     |      height
-------------+------------------
 Koniaków    |           742.25
 Jaworzynka  | 629.849975585938
 Istebna     | 661.260009765625
(3 rows)
```

Here, `(geomval).val` syntax is used, as it is composite data type, and in our case we don't need feature geometry.

This function is very computing expensive, as it converts raster coverage with `DumpAsPolygons`, and intersection is checked in vector space. For that reason, we should restrict raster extent or even clip it to vector AOI envelope. In my case, it was 51 seconds for relative small DEM tile. How to do it? As mentioned earlier, `ST_Clip()` comes to the rescue. Take a look at this:

```
SELECT ST_Clip(rast::raster, 1, (SELECT geometry FROM raster_ops.admin
LIMIT 1),true) FROM eudem.dem d;
```

The query resulted in 80 tiles, 79 blank, and 43 seconds execution time. For sure, we must use some spatial extent filtering. Most query execution time effective is `&&` (bounding box).

```
SELECT ST_Clip(rast::raster, 1, (SELECT geometry FROM raster_ops.admin
LIMIT 1),true) FROM eudem.dem d  WHERE (SELECT geometry FROM
raster_ops.admin) && d.rast;
```

Better, 1 row and 0.8s execution. Let's look into resulting clipped raster. But, if you look at the following figure, you will see that everything outside of admin polygon is filled with NODATA.

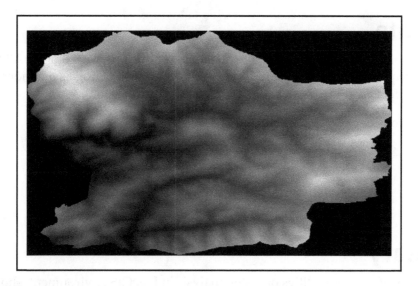

```
SELECT ST_Clip(rast::raster, 1, (SELECT ST_Envelope(geometry) FROM
raster_ops.admin LIMIT 1),true)
    FROM eudem.dem d
    WHERE (SELECT geometry FROM raster_ops.admin) && d.rast;
```

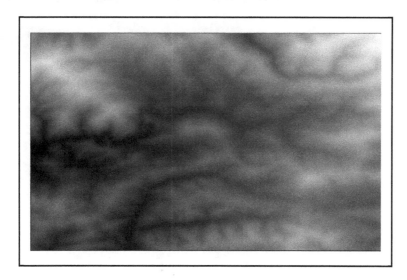

When we use ST_Envelope, not a natural border polygon, execution time is even better (0.5s in our case).

Metadata

Last but not least, let's try to get out some metadata for our raster coverages. All the information required for it is stored in the `raster_columns` catalog as mentioned before. But PostGIS offers a variety of functions for metadata return.

For example, consider `ST_ScaleX`:

```
SELECT ST_ScaleX(st_slope) FROM raster_ops.slope;
 rid | st_scalex
-----+---------------
   1 |           0.0002777777778
```

Summary

That's only few of the possibilities of raster analysis in PostGIS environment, showcased in this chapter. But if you're familiar with geometry, raster should be your friend, as it speaks the same language.

Next, we'll look into exporting our dataset from PostGIS to other GIS formats, rasters too!

5
Exporting Spatial Data

By now you have gained experience importing and working with spatial data, so it is now time to get it out of a database. This chapter is all about exporting data from PostgreSQL/PostGIS to files or other data sources. Sharing data via the web is no less important, but it has its own specific process, and is taken care of in a separate chapter.

There may be different reasons for having to export data from a database, but certainly sharing it with others is among the most popular ones. Backing the data up or transferring it to other software packages for further processing are other common reasons for learning export techniques.

In this chapter, we'll have a closer look at the following:

- Exporting data using `\COPY` (and `COPY`)
- Exporting vector data using `pgsql2shp`
- Exporting vector data using `ogr2ogr`
- Exporting data using GIS clients
- Outputting rasters using GDAL
- Outputting rasters using psql
- Using the PostgreSQL backup functionality

Basically, this chapter is structured in a very similar way to Chapter 1, *Importing Spatial Data*, which dealt with importing the data. We just do the steps the other way round. In other words, this chapter may give you a bit of a déjà vu feeling.

Exporting data using \COPY in psql

When we were importing data, we used the psql \COPY FROM command to copy data from a file to a table. This time, we'll do it the other way round - from a table to a file - using the \COPY TO command.

\COPY TO can not only copy a full table, but also the results of a SELECT query, and that means we can actually output filtered sub datasets of the source tables.

Similarly to the method we used to import, we can execute \COPY or COPY in different scenarios: We'll use psql in interactive and non-interactive mode, and we'll also do the very same thing in PgAdmin.

It is worth remembering that COPY can only read/write files that can be accessed by an instance of the server, usually files that reside on the same machine as the database server.

For detailed information on \COPY syntax and parameters, type:

```
\h copy
```

Exporting data in psql interactively

In order to export the data in interactive mode, we first need to connect to the database using psql:

```
psql -h localhost -p 5434 -U postgres
```

Then type the following:

```
\c mastering_postgis
```

Once connected, we can execute a simple command:

```
\copy data_import.earthquakes_csv TO earthquakes.csv WITH DELIMITER ';' CSV HEADER
```

The preceding command exported a data_import. earthquakes_csv table to a file named earthquakes.csv, with ';' as a column separator. A header in the form of column names has also been added to the beginning of the file. The output should be similar to the following:

```
COPY 50
```

Basically, the database told us how many records have been exported. The content of the exported file should exactly resemble the content of the table we exported from:

```
time;latitude;longitude;depth;mag;magtype;nst;gap;dmin;rms;net;id;updated;p
lace;type;horizontalerror;deptherror;magerror;magnst;status;locationsource;
magsource
2016-10-08
14:08:08.71+02;36.3902;-96.9601;5;2.9;mb_lg;;60;0.029;0.52;us;us20007csd;20
16-10-08 14:27:58.372+02;15km WNW of Pawnee,
Oklahoma;earthquake;1.3;1.9;0.1;26;reviewed;us;us
```

As mentioned, \COPY can also output the results of a SELECT query. This means we can tailor the output to very specific needs, as required. In the next example, we'll export data from a spatialized earthquakes table, but the geometry will be converted to a **WKT (wellknown text)** representation. We'll also export only a subset of columns:

```
\copy (select id, ST_AsText(geom) FROM
data_import.earthquakes_subset_with_geom) TO earthquakes_subset.csv WITH
CSV DELIMITER '|' FORCE QUOTE *  HEADER
```

Once again, the output just specifies the amount of records exported:

```
COPY 50
```

The executed command exported only the id column and a WKT-encoded geometry column. The export force wrapped the data into quote symbols, with a pipe (|) symbol used as a delimiter. The file has header:

```
id|st_astext
"us20007csd"|"POINT(-96.9601 36.3902)"
"us20007csa"|"POINT(-98.7058 36.4314)"
```

Exporting data in psql non-interactively

If you're still in psql, you can execute a script by simply typing the following:

```
\i path/to/the/script.sql
```

For example:

```
\i code/psql_export.sql
```

The output will not surprise us, as it will simply state the number of records that were outputted:

```
COPY 50
```

If you happen to have already quitted psql, the cmd \i equivalent is −f, so the command should look like this:

```
Psql −h localhost −p 5434 −U postgres −d mastering_postgis −f
code/psql_export.sql
```

Not surprisingly, the cmd output is once again the following:

```
COPY 50
```

Exporting data in PgAdmin

In PgAdmin, the command is COPY rather than \COPY. The rest of the code remains the same. Another difference is that we need to use an absolute path, while in psql we can use paths relative to the directory we started psql in.

So the first psql query translated to the PgAdmin SQL version looks like this:

```
copy data_import.earthquakes_csv TO
'F:\mastering_postgis\chapter06\earthquakes.csv' WITH DELIMITER ';' CSV
HEADER
```

The second query looks like this:

```
copy (select id, ST_AsText(geom) FROM
data_import.earthquakes_subset_with_geom) TO
'F:\mastering_postgis\chapter06\earthquakes_subset.csv' WITH CSV DELIMITER
'|' FORCE QUOTE * HEADER
```

Both produce a similar output, but this time it is logged in PgAdmin's query output pane **Messages** tab:

```
Query returned successfully: 50 rows affected, 55 msec execution time.
```

 It is worth remembering that COPY is executed as part of an SQL command, so it is effectively the DB server that tries to write to a file. Therefore, it may be the case that the server is not able to access a specified directory. If your DB server is on the same machine as the directory that you are trying to write to, relaxing directory access permissions should help.

Exporting vector data using pgsql2shp

`pgsql2shp` is a command-line tool that can be used to output PostGIS data into shapefiles. Similarly to outgoing \COPY, it can either export a full table or the result of a query, so this gives us flexibility when we only need a subset of data to be outputted and we do not want to either modify the source tables or create temporary, intermediate ones.

pgsql2sph command line

In order to get some help with the tool just type the following in the console:

```
pgsql2shp
```

The general syntax for the tool is as follows:

```
pgsql2shp [<options>] <database> [<schema>.]<table>
pgsql2shp [<options>] <database> <query>
```

Shapefile is a format that is made up of a few files. The minimum set is SHP, SHX, and DBF. If PostGIS is able to determine the projection of the data, it will also export a PRJ file that will contain the SRS information, which should be understandable by the software able to consume a shapefile.

If a table does not have a geometry column, then only a DBF file that is the equivalent of the table data will be exported.

Let's export a full table first:

```
pgsql2shp -h localhost -p 5434 -u postgres -f full_earthquakes_dataset
mastering_postgis data_import.earthquakes_subset_with_geom
```

The following output should be expected:

```
Initializing...
Done (postgis major version: 2).
Output shape: Point
Dumping: X [50 rows].
```

Now let's do the same, but this time with the result of a query:

```
pgsql2shp -h localhost -p 5434 -u postgres -f full_earthquakes_dataset
mastering_postgis "select * from data_import.earthquakes_subset_with_geom
limit 1"
```

 To avoid being prompted for a password, try providing it within the command via the -P switch.

The output will be very similar to what we have already seen:

```
Initializing...
Done (postgis major version: 2).
Output shape: Point
Dumping: X [1 rows].
```

In the data we previously imported, we do not have examples that would manifest shapefile limitations. It is worth knowing about them, though. You will find a decent description at https://en.wikipedia.org/wiki/Shapefile#Limitations. The most important ones are as follows:

- **Column name length limit**: The shapefile can only handle column names with a maximum length of 10 characters; pgsql2shp will not produce duplicate columns, though if there were column names that would result in duplicates when truncated, then the tool will add a sequence number.
- **Maximum field length**: The maximum field length is 255; psql will simply truncate the data upon exporting.

In order to demonstrate the preceding limitations, let's quickly create a test PostGIS dataset:

Create a schema if an export does not exist:

```
CREATE SCHEMA IF NOT EXISTS data_export;
CREATE TABLE IF NOT EXISTS data_export.bad_bad_shp (
  id character varying,
  "time" timestamp with time zone,
  depth numeric,
  mag numeric,
  very_very_very_long_column_that_holds_magtype character varying,
  very_very_very_long_column_that_holds_place character varying,
  geom geometry);
INSERT INTO data_export.bad_bad_shp select * from
data_import.earthquakes_subset_with_geom limit 1;
UPDATE data_export.bad_bad_shp
```

```
SET very_very_very_long_column_that_holds_magtype = 'Lorem ipsum dolor sit
amet, consectetur adipiscing elit. Fusce id mauris eget arcu imperdiet
tristique eu sed est. Quisque suscipit risus eu ante vestibulum hendrerit
ut sed nulla. Nulla sit amet turpis ipsum. Curabitur nisi ante, luctus nec
dignissim ut, imperdiet id tortor. In egestas, tortor ac condimentum
sollicitudin, nisi lacus porttitor nibh, a tempus ex tellus in ligula.
Donec pharetra laoreet finibus. Donec semper aliquet fringilla. Etiam
faucibus felis ac neque facilisis vestibulum. Vivamus scelerisque at neque
vel tincidunt. Phasellus gravida, ipsum vulputate dignissim laoreet, augue
lacus congue diam, at tempus augue dolor vitae elit.';
```

Having prepared a vigilante dataset, let's now export it to SHP to see if our SHP warnings were right:

```
pgsql2shp -h localhost -p 5434 -u postgres -f bad_bad_shp mastering_postgis
data_export.bad_bad_shp
```

When you now open the exported shapefile in a GIS client of your choice, you will see our very, very long column names renamed to VERY_VERY_ and VERY_VE_01. The content of the very_very_very_long_column_that_holds_magtype field has also been truncated to 255 characters, and is now Lorem ipsum dolor sit amet, consectetur adipiscing elit. Fusce id mauris eget arcu imperdiet tristique eu sed est. Quisque suscipit risus eu ante vestibulum hendrerit ut sed nulla. Nulla sit amet turpis ipsum. Curabitur nisi ante, luctus nec dignissim ut.

For the sake of completeness, we'll also export a table without geometry, so we can be certain that pgsql2shp exports only a DBF file:

```
    pgsql2shp -h localhost -p 5434 -u postgres -f a_lonely_dbf
mastering_postgis "select id, place from
data_import.earthquakes_subset_with_geom limit 1"
```

pgsql2shp gui

We have already seen the PgAdmin's GUI for importing shapefiles. As you surely remember, the pgsql2shp GUI also has an **Export** tab.

 If you happen to encounter difficulties locating the pgsql2shp GUI in pgAdmin 4, try calling it from the shell/command line by executing shp2pgsql-gui. If it is not recognized, try to locate the utility in your DB directory under bin/postgisgui/shp2pgsql-gui.exe.

In order to export a shapefile from PostGIS, go to the `Plugins\PostGIS` shapefile and DBF loader 2.2 (version may vary); then you have to switch to the **Export** tab:

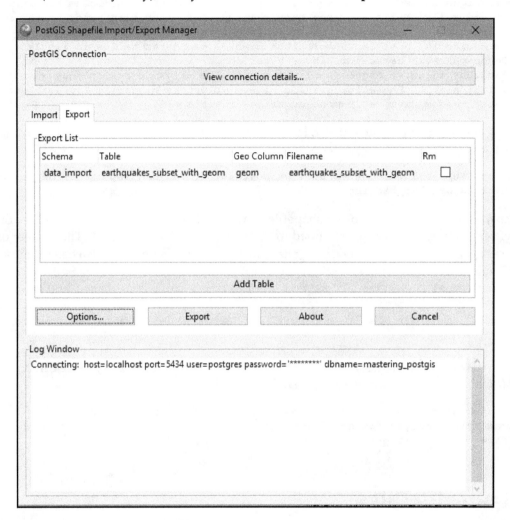

It is worth mentioning that you have some options to choose from when exporting. They are rather self-explanatory:

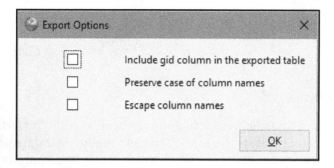

When you press the **Export** button, you can choose the output destination. The log is displayed in the **Log Window** area of the exporter GUI:

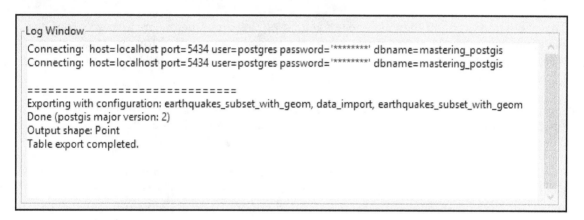

Exporting vector data using ogr2ogr

We have already seen a little preview of `ogr2ogr` exporting the data when we made sure that our KML import had actually brought in the proper data. This time we'll expand on the subject a bit and also export a few more formats, to give you an idea of how sound a tool `ogr2ogr` is.

In order to get some information on the tool, simply type the following in the console:

```
ogr2ogr
```

Alternatively, if you would like to get some more descriptive info, visit `http://www.gdal.org/ogr2ogr.html`.

You could also type the following:

```
ogr2ogr –long–usage
```

The nice thing about `ogr2ogr` is that the tool is very flexible and offers some options that allow us to export exactly what we are after. You can specify what data you would like to select by specifying the columns in a `–select` parameter. `–where` parameter lets you specify the filtering for your dataset in case you want to output only a subset of data. Should you require more sophisticated output preparation logic, you can use an `–sql` parameter.

This is obviously not all there is. The usual gdal/ogr2ogr parameters are available too. You can reproject the data on the fly using the `–t_srs` parameter, and if, for some reason, the SRS of your data has not been clearly defined, you can use `–s_srs` to instruct ogr2ogr what the source coordinate system is for the dataset being processed.

There are obviously advanced options too. Should you wish to clip your dataset to a specified bounding box, polygon, or coordinate system, have a look at the `–clipsrc`, and `–clipdst` parameters, and their variations.

The last important parameter to know is `–dsco`-dataset creation options. It accepts values in the form of `NAME=VALUE`. When you want to pass more than one option this way, simply repeat the parameter. The actual dataset creation options depend on the format used, so it is advised that you consult the appropriate format information pages available via the `ogr2pgr` website.

Exporting KML revisited

You may remember that the last time we exported KML, our focus was to only get the data out of the database so we can check whether our import had actually been successful. This time we'll issue a similar command, but with one important change - we'll specify the name field and the description field for the output KML.

A full format information page can be found at
`http://www.gdal.org/drv_kml.html`.

```
ogr2ogr -f "KML" earthquakes_from_postgis.kml PG:"host=localhost port=5434
user=postgres dbname=mastering_postgis" -t_srs EPSG:4326 -dsco NameField=id
-dsco DescriptionField=place data_import.earthquakes_subset_with_geom
```

If you opened the exported KML, this time you will have noticed that the object icons have their names displayed, and when a pop-up is opened, the name and description now come from the columns specified:

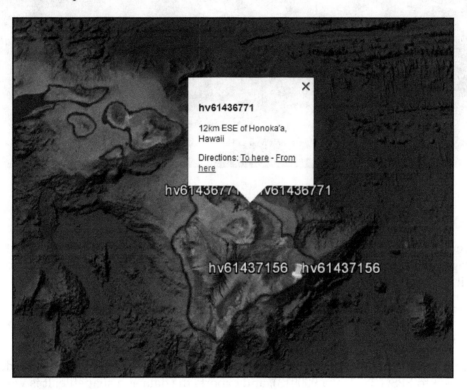

If an output format supports multiple layers, and KML does indeed, it is possible to export multiple data tables to one output file - for example:

```
ogr2ogr -f "KML" greenwich.kml PG:"host=localhost port=5434 user=postgres
dbname=mastering_postgis" -s_srs EPSG:900913 -t_srs EPSG:4326 -dsco
NameField=name -dsco DescriptionField=tags public.planet_osm_line
public.planet_osm_point
```

Our Greenwich KML should look like this, when opened in Google Earth:

Exporting SHP

We have already learned how to export SHP from PostGIS. We'll use SHP as the output format again, this time to show how to export a subset of data using ogr2ogr and how to re-project it during the export:

```
ogr2ogr -f "ESRI Shapefile" ne_coastline_islands PG:"host=localhost
port=5434 user=postgres dbname=mastering_postgis" -t_srs EPSG:3857 -where
"scalerank=1" data_import.ne_coastline
```

The preceding command extracts the islands off the 110M Natural Earth Coastline dataset, transforms it to EPSG 3857 (aka 900913), and exports the data as a shapefile. It is worth noting that because the shapefile is a format that is made up of a few files, the files were exported to a directory named ne_coastline_islands, and the file names are actually named after a schema and table we exported from, in this case data_import.ne_coastline.*. If the output file name was specified as ne_coastline_islands.shp, ogr2ogr would not create a directory for us and the exported files named ne_coastline_islands.*.

SHP is the default output format of `ogr2ogr`, so there is no need to specify it explicitly.

Let's rewrite the preceding command a bit using the `-select` parameter, so we can see it in action:

```
ogr2ogr ne_coastline_with_select_param PG:"host=localhost port=5434
user=postgres dbname=mastering_postgis" -select "scalerank,featurecla" -
t_srs EPSG:3857 -where "scalerank=1" data_import.ne_coastline
```

The result should be exactly the same in terms of exported data; we just got there using a bit of a different path, this time specifying the columns explicitly.

Similarly to `pgsql2shp`, if there is no geometry to output, `ogr2ogr` will only output a DBF when exporting to SHP :

```
ogr2ogr -f "ESRI Shapefile" ne_coastline_data_only_with_sql_param
PG:"host=localhost port=5434 user=postgres dbname=mastering_postgis" -sql
"SELECT gid, scalerank, featurecla FROM data_import.ne_coastline;"
data_import.ne_coastline
```

This time, `ogr2ogr` has also created an output directory for us. The DBF file name is not that obvious initially, but it makes perfect sense: `sql_statement`. You have surely noticed the presence of the `-sql` parameter that was used to customize the output content.

Exporting MapInfo TAB and MIF

Once you have exported a few datasets using `ogr2ogr`, you can see that exporting to other formats is a very similar process. The most important thing is to remember to consult the data set creation options for the format-specific options that may affect the export. In this case, we'll use the `-dsco FORMAT` in order to output the MIF in our second example. We will first deal with the TAB export (made of four files: DAT, ID, MAP, and TAB):

```
ogr2ogr -f "MapInfo File" greenwich_park_1.tab PG:"host=localhost port=5434
user=postgres dbname=mastering_postgis" public.planet_osm_line
```

Now we'll export to MIF (composed of two files: MIF and MID):

```
ogr2ogr -f "MapInfo File" greenwich_park_2.mif PG:"host=localhost port=5434
user=postgres dbname=mastering_postgis" -dsco FORMAT=MIF
public.planet_osm_point
```

Exporting to SQL Server

Exporting data from a database to a file that can be shared is surely a rather common task, and `ogr2ogr` is a great tool for it. If that is not enough, you can also easily use it to transfer data from one database to another. Here is how to transfer data from PostGIS to SQL Server:

```
ogr2ogr -f "MSSQLSpatial" MSSQL:"
server=CM_DOM\MSSQLSERVER12;database=mastering_postgis;trusted_connection=y
es;" PG:"host=localhost port=5434 user=postgres dbname=mastering_postgis" -
sql "SELECT name, ST_SetSRID(ST_Point(easting, norting), 27700) as geom
FROM data_import.osgb_poi" -s_srs EPSG:27700 -t_srs EPSG:4326 -lco
GEOM_TYPE=GEOMETRY -nln osgb_poi
```

The preceding command does a few things, so let's have a closer look at it. Basically, it transfers data from a PostGIS-enabled PostgreSQL databasen defined in the PG connection string to a Microsoft SQL Server spatial databasen defined in the MS SQL connection string.

> I am using an SQL-Server-trusted connection here - this means my OS user is recognized by the database, so I do not have to pass the credentials. In case I wanted to do it anyway, I would use the UID and PWD parameters with the appropriate data, as we saw in the *Foreign Data Wrappers* section.

Data is extracted using an SQL command and passed to a table in another database; data is also transformed from EPSG:27700 (OSGB 1936 / British National Grid) to EPSG:4326. Because a select query is used to extract the data, a `-nln` parameter allows us to specify the output table name so that it does not become `sql_statement`. I have also used a `GEOM_TYPE` layer creation option to specify that I was after the GEOMETRY type for my spatial data (GEOMETRY is the default, but I wanted to highlight this option).

> It is worth noting that the data transferred from PostGIS is not spatial - it gets spatialized in the select query by using the `ST_Point` and `ST_SetSRID` PostGIS methods.

ogr2ogr GUI

We have seen the `ogr2gui` already in the `Chapter 1`, *Importing Spatial Data*. Since its interface is self-explanatory, I will not really elaborate on it here, but instead, as a reminder, I'll provide the following screenshot:

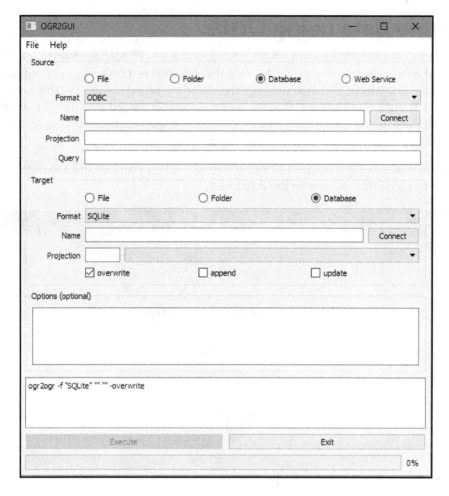

While for some the `ogr2logui` may seem a nice helper, it has some limitations, and in order to fully utilize the `ogr2ogr` goodies, one has to use it in a command-line environment.

Exporting data using GIS clients

Exporting data using GIS clients is no different to what we have seen already: simply connect to a database, read the data, and then output to a format of your choice. Once a connection can be established, the output formats are totally up to the client used.

In this section, we'll use two GIS clients:QGIS and Manifold.

Exporting data using QGIS

In order to connect to PostGIS from QGIS, go to Database\DB Manager and pick DB Manager. DB Manager is a very powerful tool that lets one not only simply connect and read data from a database, but also manage the database objects, such as tables, their columns, constraints, and indexes. In this scenario, we will not go into the details of how to manage a database using QGIS DB Manger, but instead we will focus on the task and simply use it to get to the data we want exported.

First, let's make sure we can connect to our database. When you expand the database node, you will see the schemas present in the database:

If you happen to not have any connections available, you may add one by using the **Add PostGIS Table(s)** tool, as shown in the following screenshot:

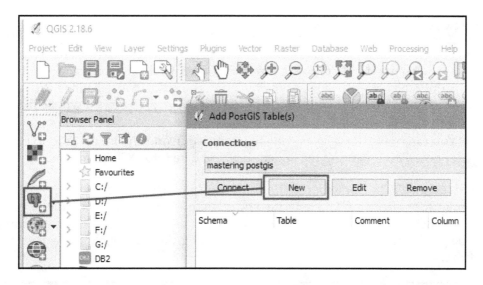

The nice thing about DB Manager is that it allows you to preview the data not only in a tabular form, but also in its spatial representation:

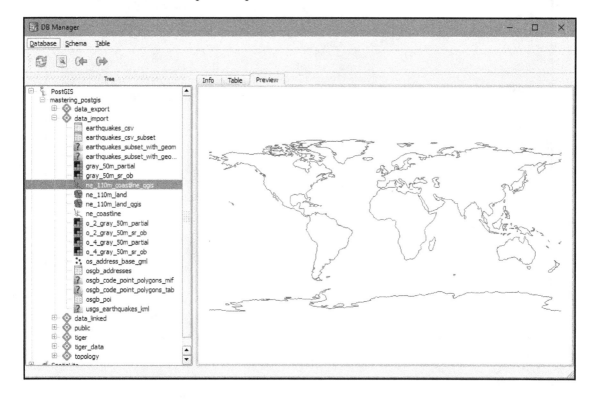

You have two options now:

- You can use the export button in the DB Manager's toolbar (the one with the arrow pointing right or to the top in the newer QGIS versions) and it will let you output the vector to SHP. At the time of writing (QGIS Las Palmas, 2.18.x), trying to export a PgRaster table to a file results in an error. I guess it will be fixed at some point.
- You can right-click a layer you are interested in and choose the Add to canvas option. This will bring the linked database data into QGIS's workspace.

Once the data is in QGIS, our export options are much, much broader. We can export to any vector format supported by QGIS to write output. In order to export vector data from QGIS, simply right-click the layer to display its context menu and choose the **Save** as option, and you will be presented with the vector data export dialog. As you see, you now have more to choose from:

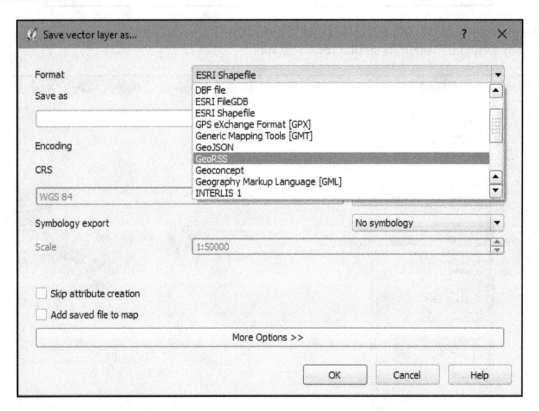

The very same approach can be applied to PgRaster tables. Simply bring the table to the QGIS workspace by right-clicking the table in the manager and choosing **Add to canvas**. Once in QGIS, our export options are way more flexible:

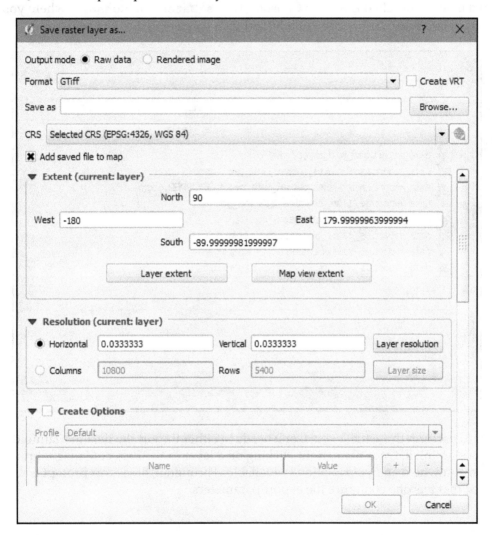

 Although not mentioned earlier, DB Manager could also be used to import the data to the PostGIS database. The process is very similar - whatever we have in the QGIS workspace that is an output table will be listed in an appropriate dropdown of DB Manager's import window.

Exporting data using Manifold.

The process of connecting to a database in Manifold is similar to the one we saw in QGIS. Manifold uses a tool called Database Console (`Tools\Database Console`) where you can define your data source connection details and then connect to the data source:

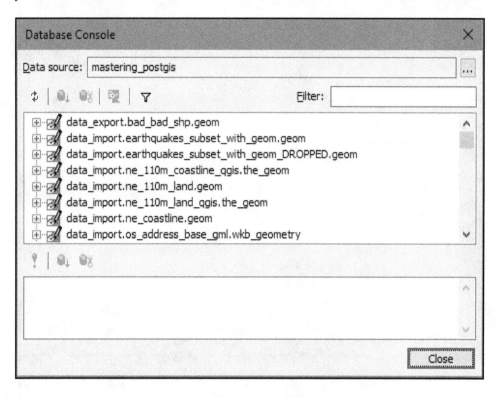

In order to perform the actual export, one needs to either import the data to Manifold's workspace or link it. Once done, right-clicking on a layer in the Project Manager brings in a context menu where an **Export** option is available. Through it, an export prompt is displayed that lets one configure the export parameters.

At the time of writing, it is not possible to consume PgRaster in Manifold, although future versions of the software are likely to happily connect to the PgRaster data too.

Outputting rasters using GDAL

We have seen `ogr2ogr` in action already, and now it is time to give two GDAL tools a spin: `gdal_translate` and `gdalwarp`. The former is a translation utility that can change the format of a raster datasource while the latter is a reprojection utility. Therefore, in order to extract a reprojected raster from a PostGIS database, we need to perform two steps:

1. Extract the raster using `gdal_translate`
2. Perform a reprojection using `gdalwarp`

In order to get some help with both utilities, simply type the following in the console:

`gdal_translate`

You could also use the following:

`gdalwarp`

The most important parameters of `gdal_translate` for our scenario are:

- `-of`: Specifies the output format (use `gdal_translate` -formats to obtain information on the supported formats).
- `-outsize xsize[%] ysize[%]`: Specifies the output raster size. It can be expressed in pixels or a percentage. If it is not used 100%, then the original size is assumed
- `PG` (PgSQL connection opts): PgSQL connection options. It is possible to pass a more complex where clause using a command line where you switch to ogrinfo.
- `where`: Filtering using a `where` clause.
- `-projwin upper left X, upper left Y, lower right X, lower right Y`: Provides a means for clipping the output raster using projected coordinates. If for some reason you prefer using pixel coordinates, use the `-srcwin` parameter.

So, let's export our Natural Earth raster to GeoTiff first, making sure that we export the top-right section and that we shrink it by half on the way:

```
gdal_translate PG:"host=localhost port=5434 user=postgres
dbname=mastering_postgis schema=data_import table=gray_50m_partial
where='filename=\'gray_50m_partial_bl.tif\'' mode=2" -of GTiff -outsize 50%
50% gray_50m_partial_small.tiff
```

You should see similar output to the following:

```
Input file size is 5400, 2700
0...10...20...30...40...50...60...70...80...90...100 - done.
```

The following is taken from the PostGIS documentation: *The mode=2 is required for tiled rasters and was added in PostGIS 2.0 and GDAL 1.8 drivers. This does not exist in GDAL 1.7 drivers.*

The `where` clause can also contain spatial filtering. The following command exports a tile that intersects with some declared bounds - in this case, it's roughly Poland's bounding box:

```
gdal_translate PG:"host=localhost port=5434 user=postgres
dbname=mastering_postgis schema=data_import table=gray_50m_partial
where='ST_Intersects(rast, ST_MakeEnvelope(14,49,24,55,4326))' mode=2" -of
GTiff -outsize 50% 50% gray_50m_partial_small.tiff
```

Although this example uses an `ST_Intersects` function, it does not clip the raster to the bounds used. Instead, it just picks a tile that intersects them. However, we still have some options to clip a raster, using, for example, the aforementioned `-projwin` parameter. To expand on the previous example, let's now crop the territory of Poland out of the larger raster:

```
gdal_translate PG:"host=localhost port=5434 user=postgres
dbname=mastering_postgis schema=data_import table=gray_50m_partial
where='ST_Intersects(rast, ST_MakeEnvelope(14,49,24,55,4326))' mode=2" -of
GTiff -projwin 14 55 24 49 poland.tiff
```

Should you require some more sophisticated filtering or better control over your output, you can use a view or a temporary table to prepare the data. Unfortunately, with GDAL we cannot issue an SQL select query as we were able to do with ogr2ogr.

Having exported a raster to a file, changing its projection is now a formality. Let's reproject our Poland raster to EPSG:2180 - Polish Coordinate System 92:

```
gdalwarp -s_srs EPSG:4326 -t_srs EPSG:2180 poland.tiff poland_2180.tiff
```

Should you need to adjust the resampling method when using gdalwarp, you can do so using the -r parameter. You could also use gdalwarp to crop a raster; use a -te xmin ymin xmax ymax to achieve this.

Outputting raster using psql

When you read through the PostGIS documentation carefully, you might find a really short section on using psql to export rasters. Basically, psql does not provide the functionality to easily output binary data, so the approach here is a bit hackish, and relies on the large object support in PostgreSQL (it is worth noting, though, that large object support is considered obsolete in the PgSQL documentation). The steps required to export a raster this way are as follows:

- Create a large object
- Output the raster data as bytea
- Open the large object and write the output bytea
- Export the large object
- Unlink it to clean the resources

First, let's make this work in interactive psql mode. Type the following in psql, once connected to our mastering_postgis database:

```
select
    loid,
    lowrite(lo_open(loid, 131072), gtiff) as bytesize
from (
    select
        lo_create(0) as loid,
        ST_AsGDALRaster(rast, 'GTiff') as gtiff
    from (
        select
            ST_Union(rast) as rast
        from
            data_import.gray_50m_partial
        where
            filename = 'gray_50m_partial_bl.tif'
    ) as combined
) as gdal_rasterised;
```

```
You should get a similar output to the following:
   loid  | bytesize
 --------+----------
 145707 | 14601978
 (1 row)
```

 If you happen to get errors mentioning the GDAL driver's loading problems, try enabling it by using SET postgis.gdal_enabled_drivers = 'ENABLE_ALL';.

Once you know your large object identifier, you can export it:

```
\lo_export 145707
'F:/mastering_postgis/chapter06/gray_50m_partial_bl_psql_interactive.tif'
```

The output will just be the following:

```
lo_export
```

Finally, we'll release the large object storage:

```
SELECT lo_unlink(145707);
```

The output should be as follows:

```
lo_unlink
-----------
         1
 (1 row)
```

I am purposefully not explaining the preceding code in detail because we will use its variation to perform the very same operation in one go. To do this, we can put the previous code into one SQL (you'll find the code in the chapter resources):

```
--Step 1: prepare the data for export
-------------------------------------
--Note: read this query starting at the inner most sql

--4. store the metadata in a temp table, so it can be used later to drive
the xport;
drop table if exists data_export.psql_export_temp_tbl;
create table data_export.psql_export_temp_tbl as
--3. open large object for writing, and write to it the GeoTiff data
prepared at lvl 2
select
    loid,
    lowrite(lo_open(loid, 131072), gtiff) as bytesize --pen large object  in
```

```
wrtie    mode, write gtiff data to it; write outputs the byte size
    --note: 131072 is a decimal value of the INV_WRITE flag hex value of
0x00020000 (see libpq-fs.h for details)
from (
    --2. create a large object, so it can be written to later and prepare
GeoTiff binary data
    select
        lo_create(0) as loid, --note the param here is 'mode' which is unused
and ignored as of pgsql 8.1 but left for backwards compatibility
        ST_AsGDALRaster(rast, 'GTiff') as gtiff
    from (
        --1. extract the gray_50m_partial_bl raster tiles and union them back
into one raster
        select
            ST_Union(rast) as rast
        from
            data_import.gray_50m_partial
        where
            filename = 'gray_50m_partial_bl.tif'
    ) as combined
) as gdal_rasterised;

--step 2: export the data
------------------------
select lo_export((select loid from data_export.psql_export_temp_tbl limit
1), 'F:\mastering_postgis\chapter06\gray_50m_partial_bl.tif');


--step 3: cleanup, cleanup, everybody cleanup...
------------------------------------------------
--release large object resources
select lo_unlink((select loid from data_export.psql_export_temp_tbl limit
1));
--drop temp table
drop table data_export.psql_export_temp_tbl;
```

The preceding code uses a temporary table to save some large object data, most importantly the large object identifier that is next used to perform the actual export and cleanup. This is because otherwise we would not be able to access the large file by its ID, and the export function would fail.

You can now execute the code in psql:

```
psql -h localhost -p 5434 -U postgres -d mastering_postgis -f
psql_export_raster.sql
```

The output should be similar to the following:

```
DROP TABLE
SELECT 1
 lo_export
-----------
          1
(1 row)


 lo_unlink
-----------
          1
(1 row)

DROP TABLE
```

Obviously, the very same code could be executed in PgAdmin too.

Use ST_GDALDrivers() to get a list of the supported drivers, should you wish to export to other formats.

We have not only done some hocus-pocus to export a raster off our PostGIS database; we have also used a PostGIS ST_AsGDALRaster function to output a raster to a desired format.

Exporting data using the PostgreSQL backup functionality

In the chapter on data importing techniques, I mentioned that the built-in PostgreSQL backup utilities can also be used as data transfer tools. It does, of course, depend on the scenario we are working on, but in many cases passing data as a database backup file may be a valid and reliable solution.

Obviously, saving a backup does not output the data in a form consumable by other software packages - only by Postgres itself.

`pg_dump` is the utility that provides the database backup functionality. In order to get some help with it, type the following in the command line:

```
pg_dump --help
```

A great thing about `pg_dump` is that it allows for great flexibility when backing up a database. One can dump a whole database, a single schema, multiple schemas, a single table, or multiple tables; it is even possible to exclude specified schemas or tables.

This approach is great, for example, for scenarios where the main database should remain within a private network and a set of automated tasks performs the database backup, file transfer, and database restore on a web-exposed database server.

A basic example of backing up a whole database could look like this:

```
pg_dump -h localhost -p 5434 -U postgres -c -F c -v -b -f
mastering_postgis.backup mastering_postgis
```

Backing up a schema is very similar:

```
pg_dump -h localhost -p 5434 -U postgres -c -F c -v -b -n data_import -f
data_import_schema.backup mastering_postgis
```

A single table is backed up like this:

```
pg_dump -h localhost -p 5434 -U postgres -t
data_import.earthquakes_subset_with_geom -c -F c -v -b -f
earthquakes_subset_with_geom.backup mastering_postgis
```

Summary

There are many ways of getting the data out of a database. Some are PostgreSQL specific, some are PostGIS specific. The point is that you can use and mix any tools you prefer. There will be scenarios where simple data extraction procedures will do just fine; some other cases will require a more specialized setup, SQL or psql, or even writing custom code in external languages. I do hope this chapter gives you a toolset you can use with confidence in your daily activities.

We have not touched upon exposing the data through web services - another chapter is fully devoted to this topic, so stay tuned.

6
ETL Using Node.js

In this chapter, we will focus on ETL operations. **ETL** stands for **Extract-Transform-Load**, and its name pretty much describes what we are about to do.

As a matter of fact, we are already familiar with loading the data, as well as extracting it; we also did some data transformations when we reprojected datasets, made flat data spatial, exported a subset of a dataset, or cropped a portion of a raster off a larger dataset. Indeed, we have done ETL already, although our approach involved some manually executed tasks, so it is easy to imagine how labor intensive and therefore time consuming our operations would become if we had to repeat them many times.

Not surprisingly, it is possible to make our lives easier with just a bit of scripting. Over the next few pages, we will define some hypothetical workflows, and then use Node.js to automate and chain the required operations.

 We could obviously use any other programming language, but because we're going to do some WebGIS-related stuff a bit later, using JavaScript seemed more natural.

The workflows presented will not necessarily be 100% real-world examples, but they will touch on some processes I tend to encounter more or less frequently.

 It is important to stress that there are some really well-equipped ETL tools out there, both commercial and open source. However, the point of this chapter is not to compete with them, but show how some relatively easy-to-use techniques can bring some more muscle to our already powerful PostGIS database.

One can think of ETL as mainly focused on local or database resources. This is, indeed, often the case, but basically any data processing that leads to the creation of a new dataset, even a simple data projection onto another model, can be safely named a transformation. And since we usually also have to read and write the data, we do the *E* and *L* of ETL anyway. ETL does not always have to indicate heavy data lifting operations. Also, the sequence does not always have to be *E->T->L*; some steps are not always implicit (for example, data extraction from a PostGIS to SHP may result in some data being truncated and therefore transformed, regardless of our intent).

We will focus on a couple of examples and go through the following steps:

- Set up Node.js
- Handshake with a database using Node.js's PgSQL client
- Retrieve and process JSON data
- Geocode address data
- Consume WFS data
- Output GeoJSON
- Output TopoJSON

Setting up Node.js

Before we move on to some specific examples, Node.js must be installed. If you happen to not have Node.js set up already, you can obtain installation instructions and all the necessary resources from `https://nodejs.org/`.

In order to verify the installation, simply type the following in the console:

```
node -v
```

At the time of writing, the **LTS (long term support)** version was 6.9.1.

 The source code of the examples presented in this chapter can be found in the appropriate chapter resources.

The Node.js examples presented in this chapter will be rather simplistic. The point of the chapter is not to create bullet-proof Node.js modules, but rather to present and discuss the ideas and code them in such a way that our code is self-explanatory and easy to read and understand.

Since we will be editing simple JS files that are just text files, you can use whatever editor you find suitable for the task. However, I suggest having a look at Visual Studio Code - it is a product from Microsoft, which may make it less approachable for some, but it is a worthy piece of software. It is an easy-to-use, cross-platform editor that not only lets you factor your code, but also has a really good debugger that is a life saver when the code written does not want to work from the beginning. Not to mention the fact that you're not limited to JavaScript, but can also code and debug in other languages, such as C#, C++, F#, Go, Python, PHP, TypeScript, and many more.

Making a simple Node.js hello world in the command line

Let's start with a simple `Hello world`. We will just log a message to the console.

First, we need to create a Node.js workspace, and this is done by executing the following in the code destination folder:

```
npm init
```

You will have to provide some further information on the module you are about to create. If you prefer `npm` populating the default values for you, just issue the following command:

```
npm init -f
```

If you choose the `-f` parameter, it will create a package and force the default settings. The package name should be taken from the folder name the package is created within.

At this stage, our Node.js package has been created, so let's have a look at the `package.json`. Some basic information on the package can be found inside this file, as well as the package entry point, which in our case is the default `index.js`.

We can now start writing the code. Let's create an `index.js` file so we have a placeholder for the code we are about to write. In this example, the code will be very simple indeed:

```
console.warn('Hello world!');
```

Once the file has been saved, we can execute it. In order to do so, simply type the following in the console:

```
node index.js
```

The output should not really surprise us:

Hello world!

 The source code for this example is available in the chapter's resources in the `code/01_hello_world` directory.

Making a simple HTTP server

Since we will use some WebGIS a bit later, let's make a web `Hello world` too. For this, a new project should be created using npm. Once ready, we will need an additional HTTP module to create a web server. This is a default Node.js module, so we just need to require it in our code:

```
const http = require('http');
const server = http.createServer((req, res) => {
    console.warn('Processing request...');
    res.end('Hello world!');
});
const port  = 8080;
server.listen(port,  () => {
    console.warn('Server listening on http://localhost:%s', port);
});
```

In the preceding code, we bring in the HTTP module via a `require` method and assign it to a variable. Next, a server is created - whenever it receives a request it logs a message to the console and replies with `Hello world!`. The final step is making the server listen on port `8080`.

When you now launch the script via `node index.js`, you should see similar output in the console:

Server listening on http://localhost:8080

Whenever you navigate to `http://localhost:8080`, you should see a `Hello world!` message in your browser and the following in the console:

Processing request...

The source code for this example is available in the chapter's resources in the `code/02_hello_world_web` directory.

Handshaking with a database using Node.js PgSQL client

Before doing some more fancy stuff, we should learn how to talk to our PostgreSQL database. In order to do so, once an appropriate project is created, we need to install a DB client. I used node-PostgreSQL:

```
npm install pg --save
```

npm is the Node.js package manager, a utility that helps managing node packages: creating them, installing remote package, and so on. You will find more information at `https://docs.npmjs.com/`.

The code should be created in an `index.js` file:

```
const pg = require('pg');

//init client with the appropriate conn details
const client = new pg.Client({
    host: 'localhost',
    port: 5434,
    user: 'postgres',
    password: 'postgres',
    database: 'mastering_postgis'
});

//connect to the db
client.connect(function(err){
    if(err){
        console.warn('Error connecting to the database: ', err.message);
        throw err;
    }

    //once connected we can now interact with a db
    client.query('SELECT PostGIS_full_version() as postgis_version;',
    function(err, result){
        if(err){
```

```
            console.warn('Error obtaining PostGIS version: ', err.message);
            throw err;
        }

        //there should be one row present provided PostGIS is installed. If
        not, executing query would throw.
        console.warn(result.rows[0].postgis_version);

        //close the connection when done
        client.end(function(err){
            if(err){
                console.warn('Error disconnecting: ', err.message);
                throw err;
            }
        });
    });
});
```

In short, the preceding code brings in a `node-postgres` package, creates a DB client with the appropriate connection details, and, once connected, it retrieves PostGIS version information and prints it to the console.

When you run this code via `node index.js`, you should see a similar output to the following:

```
POSTGIS="2.2.1 r14555" GEOS="3.5.0-CAPI-1.9.0 r4090" SFCGAL="1.2.2"
PROJ="Rel. 4.9.1, 04 March 2015" GDAL="GDAL 2.0.1, released 2015/09/15"
LIBXML="2.7.8" LIBJSON="0.12" TOPOLOGY RASTER
```

The source code for this example is available in the chapter's resources in the `code/03_db_handshake` directory.

As you may have noticed, the asynchronous nature of Node.js requires us to await callbacks or subscribe to events in order to process the results of the functions called. This quickly becomes callback hell if we need to perform some more logic, and because of that, since we will now wrap our functions into promises, we can execute the code and avoid overnesting callbacks. Consider the following code:

```
var f1 = function(p1, p2){
    return new Promise((resolve, reject)=>{
        setTimeout(()=>{
            if(!p1){
                reject('whoaaa, f1 err - no params dude!');
            }
            else {
```

```
                console.warn(`f1 processig following params: ${p1},
                ${p2}`);
                resolve({p1: 'P3', p2: 'P4'});
            }
        }, 500);
    });
}

var f2 = function(input){
    return new Promise((resolve, reject)=>{
        setTimeout(()=>{
            try {
                console.warn(`f2 params are: ${input.p1}, ${input.p2}`);
            }
            catch(err){
                reject(err.message);
            }
        }, 500);
    });
}

//f1 & f2 executed one by one
f1('p1', 'p2').then(f2);

//f1 throws, execution goes to the next catch and since the chain ends
there, execution stops
f1().then(f2).catch(err => {
    console.warn(`Uups an error occured: ${err}`);
});

//f1 throws, err is processed in the next catch, and the execution
continues to throw in f2 that is caught by the last catch
f1().catch(err => console.warn(`Uups an error occured:
${err}`)).then(f2).catch(err => console.warn(`Uups an error occured:
${err}`));
```

In the preceding code, we define two functions that do not return a potentially expected value, but a `Promise` object instead. Because of that, we can chain them together so that once one is resolved another one is called. In a scenario where a `Promise` object does not resolve, the very next `catch` function in the chain is called. Unless a `catch` function throws, the remaining <then> after `catch` is executed.

 The source code for this example is available in the chapter's resources in the `code/04_promises` directory.

While some further reading on promises may be beneficial, the basic ideas behind promises are as follows:

- A promise represents a value that can be handled at any time in the future. The consumer is guaranteed to receive that value regardless of the time a handler has been registered.
- A promise value is immutable.

As we go forward, promises will be used to simplify our code and make it more readable.

Retrieving and processing JSON data

Let's imagine our company provides data services for local governments. The company does not own or generate data; its services are mainly based on aggregating data from external sources and providing them in a digested form. One of the services offered is based on weather data, and provides early alerts when weather forecasts extend certain configured parameters.

In this example, we will simulate services for Polish local government and use the following data sources:

- The third level of administrative boundaries of Poland
- OpenWeatherMap data services

The administrative boundaries simulate our *own* data source from which we will be generating weather alerts, while **OpenWeatherMap** is a weather data service provider.

Importing shapefiles revisited

Let's first prepare our *company database*. In order to do so, we first need to obtain the boundaries of the administrative divisions of Poland. The official data can be downloaded from http://www.codgik.gov.pl/index.php/darmowe-dane/prg.html, but this dataset is quite large and requires some further processing. Instead, we will use a level three administrative boundaries data extract based on the official data and provided by the folks at GIS Support (http://www.gis-support.pl/downloads/gminy.zip).

We already know how to load shapefiles using different techniques, so let's take a little detour before we start reading the JSON data and load our shapefile using Node.js. The process will look like this:

1. Download the data.
2. Extract the files.
3. Make sure our database has an appropriate schema.
4. Import the shapefile data.

 The source code for this example is available in the chapter's resources in the `code/05_importing_shp` directory.

Once our module project has been created, we need some extra modules that will simplify our work:

```
npm install pg --save
npm install unzip --save
```

Let's code our first step:

```
/**
 * downloads a file
 */
const download = function(url, destination){
    return new Promise((resolve, reject) => {

        console.log(`Downloading ${url} to ${destination}...`);

        let file = fs.createWriteStream(destination);
        let request = http.get(url, function(response){
            response.on('data', (chunk)=>{ progressIndicator.next() });
            response.pipe(file);
            file.on('finish', () => {
                progressIndicator.reset();
                console.log("File downloaded!");
                file.close();
                resolve(destination);
            });
        }).on('error', (err)=>{
            fs.unlink(destination);
            reject(err.message);
        });
    });
}
```

This is a simple download function that downloads a specified file off the Internet and saves it under the specified file name.

 If you happen to experience problems downloading the data, you should find it under the `data/05_importing_shp` directory.

Once we have a ZIP archive locally, we need to extract it in order to import our shapefile to the database:

```
/**
 * unzips a specified file to the same directory
 */
const unzipFile = function(zipFile){
    return new Promise((resolve, reject) => {
        console.log(`Unzipping '${zipFile}'...`);

        //Note: the archive is unzipped to the directory it resides in
        fs.createReadStream(zipFile)
            .on('data', (chunk)=>{ progressIndicator.next() })
            .pipe(unzip.Extract({ path: path.dirname(zipFile) }))
            //when ready return file name, so can use it to load a file to
            the db...
            .on('close', ()=>{
                progressIndicator.reset();
                console.log('Unzipped!');
                resolve(zipFile.replace('zip', 'shp')); //Note: in this
                case shp file name is same as the archive name!
            });
    });
}
```

At this stage, we should have our shapefile extracted and ready to be imported. However, before we initialize the import, let's make sure our schema exists. For this, we will just try to create it, if does not exist:

```
/**
 * checks if database is ready for data import
 */
const dbCheckup = function(shp){
    return new Promise((resolve, reject) => {
        console.log('Checking up the database...');

        let client = new pg.Client(dbCredentials);

        client.connect((err) => {
```

```
            if(err){
                reject(err.message);
                return;
            }

            client.query(`CREATE SCHEMA IF NOT EXISTS ${schemaName};`,
            (err, result) => {
                if(err){
                    reject(err.message);
                }
                else {
                    client.end();
                    console.log('Database ready!');
                    resolve(shp);
                }
            });
        });
    });
}
```

Since we got here, it looks like we are ready to import the shapefile to the database. In order to do so, we will simply call `ogr2ogr` and pass it the parameters we are after, as we did in the data import chapter:

```
/**
 * loads a shapefile to a database
 */
const dbLoad = function(shp){
    return new Promise((resolve, reject) => {
        console.log('Loading shapefile...');
        let dbc = dbCredentials;
        let cmd = `ogr2ogr -f "PostgreSQL" PG:"host=${dbc.host}
        port=${dbc.port} user=${dbc.user} dbname=${dbc.database}" "${shp}"
        -t_srs
        EPSG:2180 -nlt PROMOTE_TO_MULTI -nln ${schemaName}.${tblName}
        -overwrite -lco GEOMETRY_NAME=geom`;

        console.log(`Executing command: ${cmd}`);

        exec(cmd, (err, stdout, stderr) => {
            if(err){
                reject(err.message);
                return;
            }
            console.log(stdout || stderr);
            resolve();
        });
    });
```

```
}
```

 I used `ogr2ogr` in this example, but in fact I could well use psql in non-interactive mode or `shp2pgsql`.

Also, you may have noticed the usage of the `-nlt PROMOTE_TO_MULTI` parameter. Our shapefile contains MultiPolygons, and since `ogr2ogr` assumes a polygon geometry type by default when importing areas, it is needed in order to avoid import errors for MultiPolygons.

At this stage, all the data should be present in the table, so a final step will be checking up on the number of imported records:

```
/**
 * counts imported records
 */
const dbLoadTest = function(){
    return new Promise((resolve, reject) => {
        console.log('Verifying import...');

        let client = new pg.Client(dbCredentials);

        client.connect((err) => {
            if(err){
                reject(err.message);
                return;
            }

            client.query(`SELECT Count(*) as rec_count FROM
            ${schemaName}.${tblName};`, (err, result) => {
                if(err){
                    reject(err.message);
                }
                else {
                    client.end();
                    console.log(`Imported ${result.rows[0].rec_count}
                    records!`);
                    resolve();
                }
            });
        });
    });
}
```

In order to execute the code we were patiently putting together, we will need to add the following to our script:

```
//chain all the stuff together
download(downloadUrl, path.join(downloadDir, fileName))
    .then(unzipFile)
    .then(dbCheckup)
    .then(dbLoad)
    .then(dbLoadTest)
    .catch(err => console.log(`uups, an error has occured: ${err}`));
```

When you execute our script, the console output should be similar to the following:

```
Downloading http://www.gis-support.pl/downloads/gminy.zip to
F:\mastering_postgis\chapter07\gminy.zip...
File downloaded!
Unzipping 'F:\mastering_postgis\chapter07\gminy.zip'...
Unzipped!
Checking up the database...
Database ready!
Loading shapefile...
Executing command: ogr2ogr -f "PostgreSQL" PG:"host=localhost port=5434
user=postgres dbname=mastering_postgis"
"F:\mastering_postgis\chapter07\gminy.shp" -t_srs EPSG:2180 -nlt
PROMOTE_TO_M
ULTI -nln weather_alerts.gminy -overwrite -geomfield geom

Verifying import...
Imported 2481 records!
```

It is not hard to imagine that our hypothetical data provider delivers some specialized data that we need to process in order to deliver value to our customers. If such a procedure happens once a month, perhaps preparing the data manually will not be a problem. If it needs to be executed more often, say once a day or every hour, the benefits of using a script and a scheduled task that executes it at a specified interval (or on demand) become immediately clear.

Consuming JSON data

Since we have our **corporate** database ready, we can focus on the defined problem. Let's try to define the steps we need to take in order to complete our task:

1. Obtain the weather forecast.
2. Process the weather data and put it into the database.

3. Assign the weather forecasts to the administrative boundaries.

4. List the administrative units that meet a hypothetical alert watch.

Our company has decided to use a weather data provider called OpenWeatherMap. One can access the data via an API, and quite a lot of information is accessible with free accounts. One can also obtain data in bulk, though this requires a paid service subscription. We are not forcing you to use a commercial account, of course; we will use some data examples that are provided free of charge so potential users can familiarize themselves with the output that is provided by the service.

In this example, we will play with a weather forecast with a 3 hour interval and a 4 day timespan. The service provides fresh data every 3 hours, so it is easy to imagine how absurd it would be to process this data manually.

An example dataset can be obtained from
`http://bulk.openweathermap.org/sample/hourly_14.json.gz.`

The source code for this example is available in the chapter's resources in the `code/06_processing_json` directory.

Let's install some node modules used by this example:

```
npm install pg --save
npm install line-by-line --save
```

We saw how to download and unzip an archive in a previous example, so these steps are skipped as there is no point in repeating them here (the source code does include the download and unzip code, though).

Our data comes as gzip, so the unzipping logic is a bit different from what we saw already; the node's zlib module is used instead.
If you happen to experience problems downloading the data, you should find it under the `data/06_processing_json` directory.

Once we have downloaded and unzipped the data, we should have a look at what's inside. Basically, each line is a JSON string with the following data (based on the actual content of the JSON):

```
{
    city: {
        coord: {
            lat: 27.716667,
            lon: 85.316666
        },
        country: "NP",
        id: 1283240
        name: "Kathmandu",
        data: [
            ...
        ],
        time: 1411447617
    }
}
```

The data property contains the array of weather forecasts we are after. The weather forecast object looks like this:

```
{
    clouds: {...},
    dt: 1411441200,
    dt_txt: '2014-09-23 03:00:00'
    main: {...},
    rain: {...},
    sys: {...},
    weather: {...},
    wind: {
        deg: 84.0077,
        speed: 0.71
    }
}
```

In our scenario, we will focus on the wind speed, which is why the other properties in the preceding code are not detailed. The wind speed is expressed in m/s. We want to alert our customers whenever the forecasted wind speed exceeds level 6 on the Beaufort scale (10.8 m/s).

We already mentioned that each line of the file is a valid JSON string. This means that we can read the data line by line, without having to load all the file content to memory.

Let's read the data for Poland first:

```
/**
 * reads weather forecast json line by line
 */
const readJson = function(jsonFile){
    return new Promise((resolve, reject) => {
        console.log(`Reading JSON data from ${jsonFile}...`);

        let recCount = 0;
        let data = [];

        //use the line reader to read the data
        let lr = new lineReader(jsonFile);

        lr.on('error', function (err) {
            reject(err.message);
        });

        lr.on('line', function (line) {

            recCount ++;

            //we're spinning through over 10k recs, so updating
            progress every 100 seems a good choice
            if(recCount % 100 === 0){
                progressIndicator.next();
            }

            //parse string to json
            var json = JSON.parse(line);

            //and extract only records for Poland
            if(json.country === 'PL'){
                data.push(json);
            }
        });

        lr.on('end', function () {
            console.warn(`Extracted ${data.length} records out of
            ${recCount}.`)
            progressIndicator.reset();
            resolve(data);
        });
    });
}
```

At this stage, we have the records prefiltered, so we're ready to load them to a database. We will load the data into two tables: one will hold the forecast pinpoint - basically this is the city we obtained the forecast for, and the other table will hold the actual forecasts per city:

```
/**
 * loads weather forecast data to database
 */
const loadData = function(data){
    return new Promise((resolve, reject) => {
        console.log('Loading data to database...');

        let client = new pg.Client(dbCredentials);

        client.connect((err) => {
            if(err){
                reject(err.message);
                return;
            }

            //prepare querries - this will hold all of the sql so we
            can execute it in one go; the content will not be strings
            though but functions to execute
            let querries = [];

            //Table setup SQL - drop (so we're clean) and (re)create
            let tableSetup = executeNonQuery(client, `DROP TABLE IF
            EXISTS ${schemaName}.${tblWeatherPoints};
            DROP TABLE IF EXISTS ${schemaName}.${tblWeatherForecasts};
            CREATE TABLE ${schemaName}.${tblWeatherPoints} (id serial
            NOT NULL, station_id numeric, name character varying, geom
            geometry);
            CREATE TABLE ${schemaName}.${tblWeatherForecasts} (id
            serial NOT NULL, station_id numeric, dt numeric, dt_txt
            character varying(19), wind_speed numeric);
            `);

            querries.push(tableSetup);

            //data preparation - query functions with params to be applied
            to the executed sql commands
            for(let d of data){
                //weather forecast point
                querries.push(
                    executeNonQuery(
                        client,
                        `INSERT INTO  ${schemaName}.${tblWeatherPoints}
                        (station_id, name, geom) VALUES($1,$2,
                        ST_Transform(ST_SetSRID(ST_Point($3, $4),
```

```
4326),2180))`,
                            [d.city.id, d.city.name, d.city.coord.lon,
                            d.city.coord.lat]
                    )
            );

            //weather forecasts
            let forecasts = [];
            let params = [];
            let pCnt = 0;
            for(let f of d.data){
                forecasts.push(`SELECT $${++pCnt}::numeric,
                $${++pCnt}::numeric, $${++pCnt}, $${++pCnt}::numeric`);
                params.push(d.city.id, f.dt, f.dt_txt, (f.wind || {})
                .speed || null);
            }

            querries.push(
                executeNonQuery(
                    client,
                    `INSERT INTO ${schemaName}.${tblWeatherForecasts}
                    (station_id, dt, dt_txt, wind_speed)
                    ${forecasts.join(' UNION ALL ')}`,
                    params
                )
            );
        }

        //finally execute all the prepared query functions and wait for
        all to finish
        Promise.all(querries)
            .then(()=> {
                client.end();
                resolve();
            })
            .catch(err=>{
                try{
                    client.end();
                }
                catch(e){}
                reject(typeof err === 'string' ? err : err.message);
            });
    });

    });
}
```

If you happened to load the data in QGIS, then at this stage, our imported datasets should look like the following:

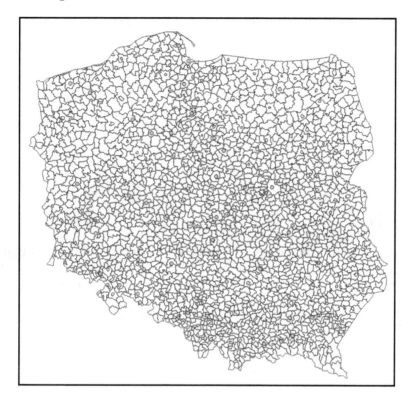

Our final step is getting the actual wind alerts. We'll do a bit more PostGIS stuff and use Node.js to execute our query. Basically, the wind speed forecasts we downloaded are not bad at all. However, there are some records with a wind speed greater than 10.8 m/s, and this is will be our cut off wind speed (wind over 10.8 falls into level 6 of the Beaufort scale and means *strong breeze*; this is when handling an umbrella becomes a challenge).

So let's think for a moment about what we have to do:

- For each administrative unit, we need to assign the nearest weather station
- We have to filter out stations with wind speed forecasts that fall into the Beaufort 6 category
- We need to select the affected administrative units

We'll initially code the query in pure SQL, as it will be much easier to digest than the same code expressed as a string in Node.js.

First, let's get a list of weather station IDs where the wind speed is forecasted to exceed our cut off point:

```sql
select
    distinct on (station_id)
    station_id,
    dt,
    dt_txt,
    wind_speed
from
    weather_alerts.weather_forecasts
where
    wind_speed > 10.8
order by
    station_id, dt;
```

The preceding query selects the weather forecasts with wind speeds greater than the mentioned 10.8 m/s and orders them by timestamp. Thanks to that, we can use distinct on `distinct on` to pick the single station IDs with the more recent forecast.

Now, let's find out the nearest weather station for each administrative unit:

```sql
select
    distinct on (adm_id)
    g.jpt_kod_je as adm_id, p.station_id, ST_Distance(g.geom, p.geom) as distance
from
    weather_alerts.gminy g, weather_alerts.weather_points p
where
    ST_DWithin(g.geom, p.geom, 200000)
order by
    adm_id, distance;
```

We use `ST_Distance` to calculate the distance between administrative units and weather stations, and then order the dataset by distance. This query gets very slow the more data is processed, so a limiting clause is used to discard weather stations that are farther than 200 km from an administrative unit (it is obvious that 200 km is way too large a range to generate sensible weather alerts, but the idea remains similar and so we will use the test data).

Finally, we need to join both queries in order to get a list of the affected administrative units:

```
select
    f.*,
    adm.*
from
    (select
            distinct on (station_id)
            station_id,
            dt,
            dt_txt,
            wind_speed
    from
            weather_alerts.weather_forecasts
    where
            wind_speed > 10.8
    order by
            station_id, dt
    ) as f
    left join (select
            distinct on (adm_id)
                    g.jpt_kod_je as adm_id, g.jpt_nazwa_ as adm_name,
                    p.station_id, p.name as station_name, ST_Distance(g.geom,
                    p.geom) as distance
            from
                    weather_alerts.gminy g, weather_alerts.weather_points p
            where
                    ST_DWithin(g.geom, p.geom, 200000)
            order by
                    adm_id, distance
    ) as adm
    on adm.station_id in (select distinct f.station_id);
```

Once our SQL is operational, we need the final piece of code, and then we should be good to go:

```
/**
 * generates wind alerts
 */
const generateAlerts = function(){
    return new Promise((resolve, reject) => {
        console.log('Generating alerts...');

        let client = new pg.Client(dbCredentials);

        client.connect((err) => {
            if(err){
```

```
                reject(err.message);
                return;
            }

            let query = `
select
    f.*,
    adm.*
from
    (select
            distinct on (station_id)
            station_id,
            dt,
            dt_txt,
            wind_speed
    from
            ${schemaName}.${tblWeatherForecasts}
    where
            wind_speed > 10.8
    order by
            station_id, dt
    ) as f

    left join (select
            distinct on (adm_id)
                g.jpt_kod_je as adm_id, g.jpt_nazwa_ as adm_name,
                p.station_id, p.name as station_name, ST_Distance(g.geom,
                p.geom) as distance
            from
                ${schemaName}.${tblAdm} g, ${schemaName}.${tblWeatherPoints}
                p
            where
                ST_DWithin(g.geom, p.geom, 200000)
            order by
                adm_id, distance
    ) as adm
    on adm.station_id in (select distinct f.station_id);`

                client.query(query, (err, result)=>{
                    if(err){
                        reject(err.message);
                    }
                    else {
                        client.end();
                        console.log(`Wind alerts generated for
                        ${result.rows.length} administrative units!`);
                        if(result.rows.length > 0){
                            let r = result.rows[0];
```

```
                    console.log(`The first one is:
                    ${JSON.stringify(r)}`);
                }
                resolve();
            }
        });
    });
});
}
```

Let's assemble the calls to the methods we have written and execute them:

```
//chain all the stuff together
download(downloadUrl, path.join(downloadDir, fileName))
    .then(gunzipFile)
    .then(readJson)
    .then(loadData)
    .then(generateAlerts)
    .catch(err => console.log(`uups, an error has occured: ${err}`));
```

The output should be similar to the following:

```
Downloading http://bulk.openweathermap.org/sample/hourly_14.json.gz to
F:\mastering_postgis\chapter07\hourly_14.json.gz...
File downloaded!
Unzipping 'F:\mastering_postgis\chapter07\hourly_14.json.gz'...
Unzipped!
Reading JSON data from F:\mastering_postgis\chapter07\hourly_14.json...
Extracted 50 records out of 12176.
Loading data to database...
Generating alerts...
Wind alerts generated for 92 administrative units!
The first one is:
{"station_id":"3081368","dt":"1411441200","dt_txt":"2014-09-23
03:00:00","wind_speed":"10.87","adm_id":"0204022","adm_name":"Jemielno","st
ation_name":"Wroclaw","distance": 53714.3452816274}
```

We have managed to transform a JSON weather forecast into a dataset with alerts for administrative units. The next steps could be exposing weather alerts via a web service, or perhaps sending out e-mails, or even SMSs.

Geocoding address data

Let's imagine that we have a potential customer database built upon yellow pages, with the customer locations expressed as addresses. I guess that using yellow pages may not be the best idea these days, but it makes a good starting point for this example.

We need to send our sales representatives to these addresses in order to establish relationships with the new customers, but first we should assign the customers to proper sales regions. Having seen St_Intersects in action already, searching for points in polygons seems like a trivial task. We need to have point geoms for this though, and we'll soon see how to go from addresses to geometry using some simple Node.js code.

Let's prepare our hypothetical **new customers** database first. We will reuse some data we have seen already, namely the Ordnance Survey GB address points. We imported this data in Chapter 1, *Importing Spatial Data*, and I assume you have not deleted the dataset already - it should be in data_import.osgb_addresses.

The records seem to be spread quite nicely, so we will simply select 100 records where we have a meaningful name and also where the building number is known:

```
--schema prepare / cleanup
create schema if not exists etl_geocoding;
drop table if exists etl_geocoding.customers;

--customers table
create table etl_geocoding.customers (
    id serial NOT NULL,
    name varchar,
    street varchar,
    street_no varchar,
    postcode varchar,
    town varchar,
    lon numeric,
    lat numeric,
    geom geometry,
    geocoded boolean
);

--get some hypothethical customers
insert into etl_geocoding.customers (
    name,
    street,
    street_no,
    postcode,
    town,
    geocoded
```

```
)

select organisation_name, thoroughfare, building_number, postcode,
post_town, false
from data_import.osgb_addresses
where organisation_name != '' and building_number != ''
limit 100;
```

Having prepared our customer database, we can now define the steps we need to take to end up with geocoded addresses:

- Extract the non-geocoded records from the database
- Use an external geocoding API in order to obtain the locations
- Pump the data back to the database

Our geocoding API will be Google Maps Geocoding API. It has its own node module, so we will be able to focus on the task without having to bother with assembling a valid URL to call the API using `http GET`. You will find more information on the Google Maps' node module at `https://github.com/googlemaps/google-maps-services-js`.

In order to use Google services, one has to generate an API key. API keys are freely available and can be created via a Google Account at `https://developers.google.com/console`.

Once our geocoding node module has been created, we will need to install some external packages:

```
npm install pg --save
npm install @google/maps --save
```

Our first step is to extract the customer records that have not yet been geocoded:

```
/**
 * reads non-geocoded customer records
 */
const readCustomers = function(){
    return new Promise((resolve, reject) => {
        console.log('Extracting customer record...');

        let client = new pg.Client(dbCredentials);

        client.connect((err) => {
            if(err){
                reject(err.message);
                return;
            }
```

```
client.query(`SELECT * FROM
${customersSchema}.${customersTable} WHERE geocoded = false
LIMIT 10;`, function(err, result){
    if(err){
        try {
            client.end();
        } catch(e){}
        reject(err.message);
        return;
    }

    client.end();

    console.log('Done!');
    resolve(result.rows);
});
});
});
}
```

 You may have noticed that I am reading only ten records at a time. This is because it lets me debug the code without having to reset the geocoding status in the database. Also, there are some usage limits with the free Google Maps account, so I am avoiding too many API calls.

Once we have the customer records at hand, we can geocode them. However, we should probably stop for a second and familiarize ourselves with the geocoder API output so that we can properly extract the information later. The following is an output example for our first address in the database: 30, GUILDHALL SHOPPING CENTRE, EX4 3HJ, EXETER:

```
{
    "results" : [
    {
        "address_components":[
            {"long_name":"30-32","short_name":"30-32",
            "types":["street_number"]},
            {"long_name":"Guildhall Shopping Centre",
            "short_name":"Guildhall Shopping Centre",
            "types":["route"]},
            {"long_name":"Exeter","short_name":"Exeter","types":
            ["locality","political"]},
            {"long_name":"Exeter","short_name":"Exeter","types":
            ["postal_town"]},
            {"long_name":"Devon","short_name":"Devon","types":
            ["administrative_area_level_2","political"]},
            {"long_name":"England","short_name":"England","types":
            ["administrative_area_level_1","political"]},
```

```
            {"long_name":"United Kingdom","short_name":"GB","types":
            ["country","political"]},
            {"long_name":"EX4 3HH","short_name":"EX4 3HH","types":
            ["postal_code"]}
        ],
        "formatted_address":"30-32 Guildhall Shopping Centre, Exeter
        EX4 3HH, UK",
        "geometry":{
            "location":{
                "lat":50.7235944,
                "lng":-3.5333662
            },
            "location_type":"ROOFTOP",
            "viewport":{
            "northeast":
            {"lat":50.7249433802915,"lng":-3.532017219708498},
            "southwest":
            {"lat":50.7222454197085,"lng":-3.534715180291502}
            }
        },
        "partial_match":true,
        "place_id":"ChIJy3ZkNDqkbUgR1WXtac_0ClE",
        "types":["street_address"]
    },
    "status" : "OK"
}
```

Once we know what the geocoder data looks like, we can easily code the geocoding procedure:

```
/**
 * generates a geocoding call
 */
const generateGeocodingCall = function(gMapsClient, customer){
    return new Promise((resolve, reject) => {
        let address = `${customer.street_no} ${customer.street},
        ${customer.postcode}, ${customer.town}`;

        gMapsClient.geocode({
          address: address
        }, (err, response) => {
          if(err){
              reject(err.message);
              return;
          }

          if(response.json.error_message){
              console.log(response.json.status,
```

```
                    response.json.error_message);
                    reject(err);
                    return;
                }

                //update customer
                let geocoded = response.json.results[0];
                if(geocoded){
                    customer.geocoded = true;
                    customer.lon = geocoded.geometry.location.lng;
                    customer.lat = geocoded.geometry.location.lat;
                }

                resolve();
            });
        });
    }
```

In order to make our geocoding call work for us, we need to call it for the retrieved records. Let's do it this way:

```
/**
 * geocodes specified customer addresses
 */
const geocodeAddresses = function(customers){
    return new Promise((resolve, reject) => {
        console.log('Geocoding addresses...');

        let gMapsClient = require('@google/maps').createClient({
            key: gMapsApiKey
        });

        //prepare geocoding calls
        let geocodingCalls = [];
        for(let c of customers){
            geocodingCalls.push(
                generateGeocodingCall(gMapsClient, c)
            );
        }

        //and execute them
        Promise.all(geocodingCalls)
            .then(()=>resolve(customers))
            .catch((err) => reject(err));
    });
}
```

At this stage, we should have our customer records geocoded so we can save them back to the database. As you may expect, this is rather straightforward:

```
/**
 * saves geocoded customers back to the database
 */
const saveCustomers = function(customers){
    return new Promise((resolve, reject) => {

        console.log('Saving geocoded customer records...');

        let client = new pg.Client(dbCredentials);

        client.connect((err) => {
            if(err){
                reject(err.message);
                return;
            }

            const updateSQLs = [];
            var pCounter = 0;

            for(let c of customers){
                updateSQLs.push(executeNonQuery(client, `UPDATE
                ${customersSchema}.${customersTable} SET
                lon=$1,lat=$2,geocoded=true WHERE id=$3;`, [c.lon,
                c.lat, c.id]));
            }

            Promise.all(updateSQLs)
                .then(() => {
                    client.end();
                    resolve();
                })
                .catch((err)=>{
                    try{
                        client.end();
                    }
                    catch(e){}
                    reject(err);
                });
        });
    });
}
```

Finally, let's call our methods in a sequence and watch the *magic* happen:

```
//chain all the stuff together
readCustomers()
    .then(geocodeAddresses)
    .then(saveCustomers)
    .catch(err => console.log(`uups, an error has occured: ${err}`));
```

Consuming WFS data

Let's imagine that we work for a utilities company that is contracted to build a piece of underground pipeline. The job is not only to do the actual construction work, the company also has to negotiate with the land owners and obtain their legal agreements for the construction work. The company GIS department has been tasked to prepare a list of parcels that will be affected by the pipeline itself and the construction work - after all, builders do need to be able to get to a place with their heavy equipment.

Our job is, therefore, to do the following:

1. Buffer the pipeline geometry with a radius of 100 m.
2. Extract the buffer geometry off the database.
3. Query a WFS service to obtain parcels that intersect with the pipeline buffer.
4. Load the parcels data to the PostGIS database.
5. Prepare a report with the parcel data.

We have visited Poland and the UK in the previous examples. For this example, we will fly over to New Zealand and consume a web feature service provided for us by LINZ (Land Information New Zealand). In order to use LINZ services, we need to register and create an API key. You can do this at `https://data.linz.govt.nz/accounts/register/`. Then, when ready, follow the API key generation instructions at `http://www.linz.govt.nz/data/linz-data-service/guides-and-documentation/creating-an-api-key`.

We have received geometry of the pipeline in question as a shapefile, so let's import it to the database. First, let's ensure that our schema is intact:

```
create schema if not exists etl_pipeline;
```

Next, we'll let `ogr2ogr` do the work for us:

```
ogr2ogr -f "PostgreSQL" PG:"host=localhost port=5434 user=postgres
dbname=mastering_postgis" "pipeline.shp" -t_srs EPSG:2193 -nln
etl_pipeline.pipeline -overwrite -lco GEOMETRY_NAME=geom
```

 The data for this example can be found in the `data/08_consumig_wfs` directory.

The pipeline is located just on the outskirts of New Plymouth, near Egmont National Park, and is marked red on the following screenshot:

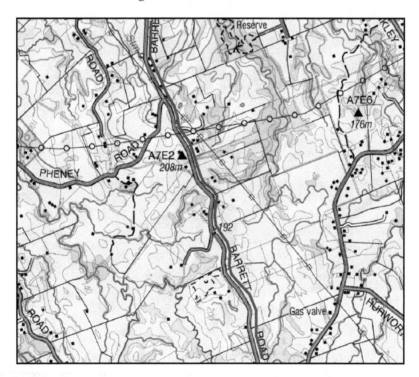

At this stage, we're ready to write the code again. We will warm up by buffering our pipeline. Let's assume that we need 5 m on each side for heavy equipment access:

```
/**
 * buffers the pipeline and returns a buffer geom as WKT
 */
const getPipelineBuffer = function(){
    return new Promise((resolve, reject) => {
```

```
        console.log('Buffering pipeline...');

        let client = new pg.Client(dbCredentials);

        client.connect((err) => {
            if(err){
                reject(err.message);
                return;
            }

            //note
            client.query(`select ST_AsGML(ST_Buffer(geom, 5, 'endcap=round
            join=round')) as gml from ${pipelineSchema}.${pipelineTable}
            limit 1;`, function(err, result){
                if(err){
                    try {
                        client.end();
                    } catch(e){}
                    reject(err.message);
                    return;
                }

                client.end();

                if(result.rows.length !== 1)
                {
                    reject('Hmm it looks like we have a little problem with
                    a pipeline...');
                }
                else {
                    console.log('Done!');
                    resolve(result.rows[0].gml);
                }
            });
        });
    });
}
```

Once we have our buffer GML ready, we can query a WFS service. We will just send a POST GetFeature request to the LINZ WFS and request the data in the same projection as our pipeline dataset, so EPSG:2193 (New Zealand Transverse Mercator 2000); since we're using JavaScript and our WFS supports JSON output, we will opt for it.

At this stage, we should have the data at hand, and since we asked for JSON output, our data should be similar to the following:

```
{
    type: "FeatureCollection",
    totalFeatures: "unknown",
    features:
    [
            {
                type: "Feature",
                id: "layer-772.4611152",
                geometry: {
                        "type": "MultiPolygon",
                        "coordinates": [...]
                },
                geometry_name: "shape",
                properties: [{
                        "id": 4611152,
                        "appellation": "Lot 2 DP 13024",
                        "affected_surveys": "DP 13024",
                        "parcel_intent": "DCDB",
                        "topology_type": "Primary",
                        "statutory_actions": null,
                        "land_district": "Taranaki",
                        "titles": "TNF1/130",
                        "survey_area": 202380,
                        "calc_area": 202486
                }]
            }
    ]
}
```

The geometry object is GeoJSON, so we should be able to easily make PostGIS read it. Let's do just that and put our parcels data in the database now:

```
/**
 * saves wfs json parcels to the database
 */
const saveParcels = function(data){
    return new Promise((resolve, reject) => {

        console.log('Saving parcels...');

        let client = new pg.Client(dbCredentials);

        client.connect((err) => {
            if(err){
                reject(err.message);
```

```
            return;
        }

        const sql = [
            executeNonQuery(client, `DROP TABLE IF EXISTS
            ${pipelineSchema}.${pipelineParcels};`),
            executeNonQuery(
                client,
                `CREATE TABLE ${pipelineSchema}.${pipelineParcels}
                (id numeric, appellation varchar, affected_surveys
                varchar, parcel_intent varchar, topology_type varchar,
                statutory_actions varchar, land_district varchar,
                titles varchar, survey_area numeric, geom geometry);`
            )
        ];

        for(let f of data.features){
            sql.push(
                executeNonQuery(
                    client,
                    `INSERT INTO ${pipelineSchema}.${pipelineParcels}
                    (id, appellation, affected_surveys, parcel_intent,
                    topology_type, statutory_actions, land_district,
                    titles, survey_area, geom)
                    VALUES
                    ($1,$2,$3,$4,$5,$6,$7,$8,$9,ST_SetSRID
                    (ST_GeomFromGeoJSON($10),2193));`,
                    [
                        f.properties.id,
                        f.properties.appellation,
                        f.properties.affected_surveys,
                        f.properties.parcel_intent,
                        f.properties.topology_type,
                        f.properties.statutory_actions,
                        f.properties.land_district,
                        f.properties.titles,
                        f.properties.survey_area,
                        JSON.stringify(f.geometry)
                    ]
                )
            );
        }

        Promise.all(sql)
            .then(() => {
                client.end();
                console.log('Done!');
                resolve();
```

```
            })
            .catch((err) => {
                client.end();
                reject(err)
            });
        });
    });
}
```

Finally, let's chain our ops:

```
//chain all the stuff together
getPipelineBuffer()
    .then(getParcels)
    .then(saveParcels)
    .catch(err => console.log(`uups, an error has occured: ${err}`));
```

Voila! We have just obtained a set of parcels that intersect with our 5 m buffer around the pipeline. We can pass the parcels information so that our legal department obtains detailed information for further negotiations. Our parcels map now looks like this:

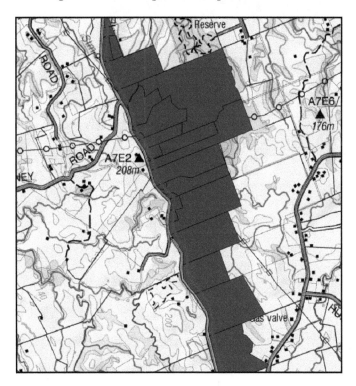

Summary

Most of the examples we have seen in this chapter involve extracting remote web-based data sources rather than processing local files. This is because there are more and more datasets provided by authorities via web services. I also find mixing remote and local resources more interesting than simply changing the formats of files or reprojecting them. As well as working with web-based data sources, we also worked with files. We downloaded ZIP and gzip archives and extracted them. We also read JSON files line by line and even made Node.js use `ogr2ogr` to import some shapefiles. We could do more file processing in `ogr2ogr`, GDAL, or psql but that would seem a bit dull.

Obviously our ETL examples were not very complex, and we did not design and execute any sophisticated data processing workflows. The important thing is that we did some task automation and have shown that adding value to our data does not have to be difficult. I do hope that, thanks to this chapter, repeatable, tedious, and time-consuming tasks are not a problem anymore; they can even be fun when defining and coding a workflow.

7
PostGIS – Creating Simple WebGIS Applications

Having learned how to use PostGIS to derive added value from our spatial data, it is now time to explore ways of sharing our spatial assets with the outside world. In this chapter, we will focus on consuming PostGIS data in WebGIS applications. In order to expose the data, we will have a look at GeoServer, but we will also write some simple web services in Node.js.

In this chapter, we will have a look at:

- Outputting vector data as web services in GeoServer:
    - Outputting vector data as WMS services in GeoServer
    - Outputting raster data as WMS services in GeoServer
    - Outputting vector data as WFS services
- Consuming WMS in ol3
- Consuming WMS in Leaflet
- Outputting and consuming GeoJSON
- Outputting and consuming TopoJSON
- Consuming WFS in ol3
- Implementing a simple CRUD application that demonstrates vector editing via web interfaces

All the WebGIS examples presented in this chapter are purposefully minimalistic. The point is not to create a fully featured web application, but rather to focus on the bits and pieces that clearly describe the topics presented and can be easily reused in other applications.

The UI library used for the examples is ExtJS - a **RIA (Rich Interface Application)** SDK available in both open source and commercial license flavors. You can obtain an open source license from here: `https://www.sencha.com/legal/GPL/`. An archive with a GPL version of the ExtJS 6.2 framework is also available in the chapter resources.

Code examples are meant to work straight away, so you should be able to simply drop the code into your web server directory and it should work without further setup. If you do not have the resources to use a web server, you can still run the examples by using the Sencha CMD tool. The ExtJS `Hello World` example describes how to use Sencha CMD without using a web server.

 There is a rather well known open source library called GeoExt that extends ExtJs with a set of web GIS-related widgets based on OpenLayers. If you happen to like ExtJs, you can find more information on GeoExt here: `https://geoext.github.io/geoext3/`.

ExtJS says Hello World

This is a short example aimed at making you comfortable with setting up and running the web application examples. First, make sure you have Sencha CMD installed. If not, you may get it from here: `https://www.sencha.com/products/extjs/cmd-download/`.

Next, you will need to set up your workspace by simply extracting the ExtJS library to `code/webgis_examples/ext`. When ready, your `ext` folder contents should look like this:

Name	Date modified	Type	Size
.sencha	31-Aug-16 15:42	File folder	
build	31-Aug-16 15:50	File folder	
classic	31-Aug-16 11:03	File folder	
cmd	31-Aug-16 15:43	File folder	
examples	31-Aug-16 15:49	File folder	
licenses	31-Aug-16 15:49	File folder	
modern	31-Aug-16 11:03	File folder	
packages	31-Aug-16 11:03	File folder	
resources	31-Aug-16 11:03	File folder	
sass	31-Aug-16 15:49	File folder	
templates	31-Aug-16 11:03	File folder	
welcome	31-Aug-16 15:50	File folder	
build.xml	31-Aug-16 11:03	XML File	24 KB
ext-bootstrap.js	31-Aug-16 11:03	JetBrains WebStorm	3 KB
index.html	31-Aug-16 11:03	HTML File	1 KB
LICENSE	31-Aug-16 15:49	File	3 KB
package.json	31-Aug-16 15:49	JSON File	2 KB
Readme.md	31-Aug-16 11:03	Markdown Source...	1 KB
release-notes.html	31-Aug-16 11:42	HTML File	827 KB
version.properties	31-Aug-16 12:02	Properties Source ...	1 KB

All the examples are located in the code directory in this chapter's resources. In order to launch our `Hello World` example, simply navigate to `code/webgis_examples/apps/01_hello_world` in your CMD and then run the following command:

```
sencha app watch
```

When you navigate to `http://localhost:1841/apps/01_hello_world/`, you should see our `Hello World` application (this is a default empty app that Sencha CMD generates):

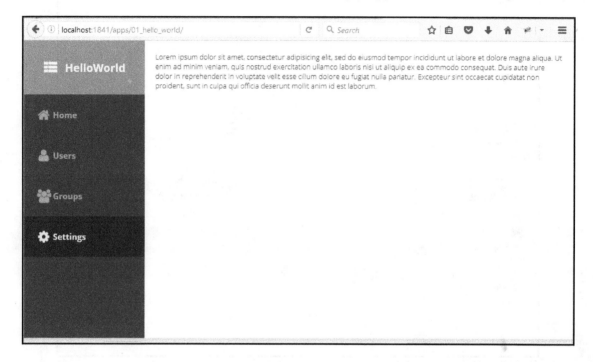

If you feel interested in learning a bit more about ExtJS, you should check out the following resources:

- `https://www.packtpub.com/web-development/mastering-ext-js-second-edition`
- `https://www.packtpub.com/web-development/ext-js-6-example`

Configuring GeoServer web services

Before we move on to writing our WebGIS code, we need to configure a service so we can expose spatial data for a web client to consume.

Our examples are based on **GeoServer** as it is considered a reference implementation of OGC web standards. Apart from that, GeoServer provides a rather intuitive administration GUI to simplify many server maintenance tasks.

If you happen to not have GeoServer installed, please get it up and running first - you can download the software from here: `http://geoserver.org/release/stable/`. GeoServer documentation may be obtained from here: `http://docs.geoserver.org/`.

Installation is rather straightforward and we suggest for the sake of simplicity you use the installation pack wrapped with the Jetty container.

While data service configuration is specific to GeoServer, you can use any alternative GeoServer of your choice, for example, MapServer (`http://www.mapserver.org/`). If you decide to not use GeoServer and go for an alternative, our code examples should not be affected (although perhaps some service URL defaults may differ).

Once you have your installation up and running on the default port 8080, when you navigate to `http://localhost:8080/geoserver` you should be greeted with the following screen:

Now as our GeoServer instance is up and running, we can move on to configuring some data sources. In order to configure GeoServer, one has to log on first - the default administrator user name is `admin` and the default password is `geoserver`.

> If you feel like learning a bit more about GeoServer- related topics, you should check out the following resources:
> `https://www.packtpub.com/application-development/geoserver-beg`
> `inner%E2%80%99s-guide`
> `https://www.packtpub.com/networking-and-servers/mastering-geos`
> `erver`
> `https://www.packtpub.com/application-development/geoserver-coo`
> `kbook.`

Importing test data

We need some data to work with before we can set up services. We have imported a fair amount of data already, but for the sake of keeping things in order, let's create a WebGIS schema and import some data:

- `http://www.naturalearthdata.com/http//www.naturalearthdata.com/downl oad/10m/raster/NE2_HR_LC_SR_W_DR.zip` - do not import this one yet!

- `http://www.naturalearthdata.com/http//www.naturalearthdata.com/downloa d/50m/raster/NE2_50M_SR_W.zip` - do not import this one yet!

- `http://www.naturalearthdata.com/http//www.naturalearthdata.com/downloa d/10m/physical/ne_10m_coastline.zip` - import into the `ne_coastline` table.

- `http://www.naturalearthdata.com/http//www.naturalearthdata.com/downloa d/10m/physical/ne_10m_reefs.zip` - import into the `ne_reefs` table.

> This time I will skip the vector import part and treat it as a little reminder exercise.
> Avoid importing the raster yet though. We will import it a bit later, so we can adjust to some GeoServer plugin requirements and use the provided tooling in order to avoid manual adjustment.

Outputting vector data as WMS services in GeoServer

 All the main *access points* of GeoServer configurations are located on the left-hand side.

We will start with configuring a workspace for the data sources we are about to expose. A workspace can be treated as a form of data encapsulation, so it is possible to keep the data in order.

In order to create a workspace, click **Workspaces** under the **Data** section. Once in the **Workspaces** section, click the **Add new workspace** link. You will be presented with a form to fill in, so simply type mastering_postgis in the **Name** and **Namespace URI** fields, and then click **Submit**.

In order to expose data from a PostgreSQL database, we need to configure a datasource first. In order to do so, navigate to the **Stores** menu in the **Data** section.

Once in the **Stores** section, click the **Add new Store** link and then choose the **PostGIS** link. You will be presented with a data source configuration form, where you will provide a database connection string as well as some other settings:

- For the workspace, pick the newly created mastering_postgis workspace
- Name your data source webgis
- Tick **Expose primary keys**, and make sure to connect to the webgis schema

Once you submit the form, provided all the parameters were valid, you should be presented with the layer creation view:

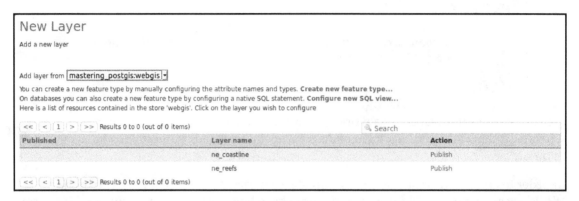

Instead of creating a layer straight away, we will do it the *harder* way so the process is described in detail and can be repeated without having to create a data store first. In order to expose a PostGIS table as a WMS service, navigate to the **Layers** menu in the **Data** section.

Once in the **Layers** section, click the **Add a new layer** link; you will be asked to choose a store - pick **mastering_postgis:webgis**. You should be presented with the **New Layer** screen that we have already seen.

Click **Publish link** next to the name of the table you are about to publish; in this scenario, let's publish the ne_coastline table. You will be presented with a **Layer creation** screen, where you should make sure to:

- Data tab:
    - Set the declared SRS to EPSG:4326
    - Set SRS handling to Force declared
    - Compute Bounding Boxes using the appropriate links
- Publishing tab:
    - Pick the default style for a line - it is a blueish line

Once the previous things have been set, save the layer by clicking the **Save** button. If all the settings were correct, you should be presented with the **Layers screen**, and a newly created layer should be present in the list of configured layers.

Let's make sure our layer has been defined properly. In order to do so, navigate to **Layer Preview** in the **Data** section, locate the layer, and click the link that opens a preview in OpenLayers. You should see a screen similar to the following:

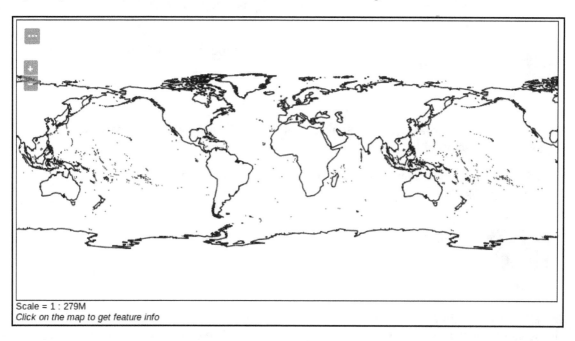

Scale = 1 : 279M
Click on the map to get feature info

Feel free to play a bit with the layer preview; especially take a closer look at some options available to you when you click the **triple dot** button. You should see a simple interface that allows you to change WMS layer configurations such as: WMS version, single tile/tiled, format, and antialiasing; you can even use CQL to query the dataset. When you click a feature, you will see some feature - related data.

I strongly encourage you to have a closer look at the requests issued by the browser in order to fetch WMS data. There are many tools to do this, but since we are about to write some WebGIS examples, I suggest using either the Chrome console, or if you happen to prefer Firefox, Firebug will be the choice. For example, the following URL extracts a tile with Northeast Australia, Papua New Guinea, and the northern part of the Solomon Islands:

```
http://localhost:8080/geoserver/mastering_postgis/wms?SERVICE=WMS&VERSI
ON=1.1.1&REQUEST=GetMap&FORMAT=image%2Fpng&TRANSPARENT=true&tiled=true&
STYLES&LAYERS=mastering_postgis%3Ane_coastline&tilesOrigin=-180%2C-90&W
IDTH=256&HEIGHT=256&SRS=EPSG%3A4326&BBOX=135%2C-22.5%2C157.5%2C0
```

Outputting raster data as WMS services in GeoServer

Unfortunately, PgRaster support is still not common and GeoServer does not support consuming PgRaster yet. This section shows how to use the PostgreSQL database to load a raster and then expose it via GeoServer.

 We will make use of PgRaster a bit later though - we will write a simple WMS GetMap request handler.

Exposing a GeoServer layer based on a raster stored in PostgreSQL is not as straightforward as it was with vector tables.

In order to create a raster layer, we first need to install an Image Mosaic JDBC plugin: `http://docs.geoserver.org/stable/en/user/data/raster/imagemosaicjdbc.html`.

 Do make sure to install the plugin appropriate to your GeoServer version.

Installing the plugin is rather straightforward and you should not encounter any difficulties doing so. It is just a matter of downloading and copying the content of a plugin archive (usually a JAR file) into GeoServer's `WEB-INF\lib` directory.

 In order to install the plugin, you will need to shut down your GeoServer instance and once the plugin is installed, you should bring it back.

Having installed our plugin, we now need to configure a data source for the raster layer. This is where some manual work is required. A detailed manual on how to configure GeoServer to consume raster data is here: `http://docs.geoserver.org/stable/en/user/tutorials/imagemosaic-jdbc/imagemosaic-jdbc_tutorial.html`, but since it is a bit XML-ish/SQL-ish, you find a tutorial excerpt in following section, so our configuration goes smoothly.

 All the files that we are preparing here can be found in this chapter's code directory.

Let's create some raster coverage configuration files first. We will need three XML files; create them in the GEOSERVER_DATA_DIR/coverages:

- ne_raster.postgis.xml

```xml
<?xml version="1.0" encoding="UTF-8" standalone="no"?>
<!DOCTYPE ImageMosaicJDBCConfig [
    <!ENTITY mapping PUBLIC "mapping"  "mapping.postgis.xml">
    <!ENTITY connect PUBLIC "connect"  "connect.postgis.xml">
]>
<config version="1.0">
    <coverageName name="ne_raster"/>
    <coordsys name="EPSG:4326"/>
    <scaleop  interpolation="1"/>
    <verify  cardinality="false"/>
      &mapping;
      &connect;
</config>
```

- connect.postgis.xml

```xml
<connect>
    <dstype value="DBCP"/>
    <username value="postgres"/>
    <password value="postgres"/>
    <jdbcUrl value="jdbc:postgresql://localhost:5434/mastering_postgis"/>
    <driverClassName value="org.postgresql.Driver"/>
    <maxActive value="10"/>
    <maxIdle value="0"/>
</connect>
```

- mapping.postgis.xml

```xml
<spatialExtension name="postgis"/>
<mapping>
    <masterTable name="mosaic" >
      <coverageNameAttribute name="name"/>
      <maxXAttribute name="maxx"/>
      <maxYAttribute name="maxy"/>
      <minXAttribute name="minx"/>
      <minYAttribute name="miny"/>
      <resXAttribute name="resx"/>
      <resYAttribute name="resy"/>
      <tileTableNameAtribute  name="tiletable" />
      <spatialTableNameAtribute name="spatialtable" />
    </masterTable>
    <tileTable>
```

```
        <blobAttributeName name="data" />
        <keyAttributeName name="location" />
    </tileTable>
    <spatialTable>
        <keyAttributeName name="location" />
        <geomAttributeName name="geom" />
        <tileMaxXAttribute name="maxx"/>
        <tileMaxYAttribute name="maxy"/>
        <tileMinXAttribute name="minx"/>
        <tileMinYAttribute name="minx"/>
    </spatialTable>
</mapping>
```

Once ready with the XML files, navigate to GEOSERVER_DATA_DIR/coverages, create a
ne_raster_scripts directory, and then execute the following command:

```
java -jar GEOSERVER_WEB_SERVER_WEBAPPS/geoserver/WEB-INF/lib/ gt-
imagemosaic-jdbc-16.0.jar ddl -config /
GEOSERVER_DATA_DIR/coverages/ne_raster.postgis.xml -spatialTNPrefix
ne_raster -pyramids 6 -statementDelim ";" -srs 4326 -targetDir
ne_raster_sqlscripts
```

This should output some SQL scripts
toGEOSERVER_DATA_DIR/coverages/ne_raster_scripts that will let us create the
appropriate tables for the raster data. Scripts will create tables in the public schema, so this
may be a bit inconvenient.

> In the preceding command, you should adjust some paths:
>
> - GEOSERVER_WEB_SERVER_WEBAPPS - should be your
>   GeoServer's webapps folder, where you have your GeoServer
>   instance deployed
> - GEOSERVER_DATA_DIR - GeoServer's data directory.

When you're ready, just execute the generated scripts:

- Createmeta.sql
- Add_ne_raster.sql

This should create some new tables in the public schema.

Once our database is prepared, we can take care of processing the raster. First we need to tile it using `gdal_retile`:

```
gdal_retile.py -co TFW=YES -r near -ps 256 256 -of GTiff -levels 6 -
targetDir tiles NE2_50M_SR_W.tif
```

 I am using a smaller raster here for a simple reason: a larger one would likely cause some troubles related to limited memory associated with our GeoServer's Jetty container.

With the tiles ready, we can now import them into the database. Before we do so, let's ensure that a PostgreSQL driver is located in the `lib/ext` directory of the Java runtime (you can copy the file from your `geoserver/WEB-INF/lib` directory; my Windows box had a `postgresql-9.4.1211.jar` file).

The final raster processing step is to use the `imagemosaik` plugin again to import the generated tiles into the database:

```
java -jar GEOSERVER_WEB_SERVER_WEBAPPS/geoserver/WEB-INF/lib/gt-
imagemosaic-jdbc-16.0.jar import -config
GEOSERVER_DATA_DIR/coverages/ne_raster.postgis.xml -spatialTNPrefix
ne_raster -tileTNPrefix ne_raster -dir tiles -ext tif
```

Or final step is to set up a layer in GeoServer. In order to do so, click **Stores** under the **Data** section and click the **Add new Store** link. Next, pick the `ImageMosaicJDBC` link and create a data source called `webgis_raster` under the `mastering_postgis` workspace. In the URL field, you will have to provide a file URL to the `ne_raster.postgis.xml` we created in the `GEOSERVER_DATA_DIR/coverages` directory. When you click the **Save** button, you should be taken to the **Create layer** screen:

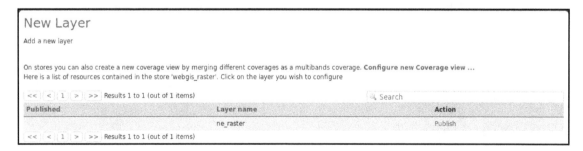

Click the **Publish** link and you will be taken to the layer configuration screen. Make sure to verify the projection information and to reload the band definitions and save the layer. Provided that you have configured the layer properly, you should now be able to preview it:

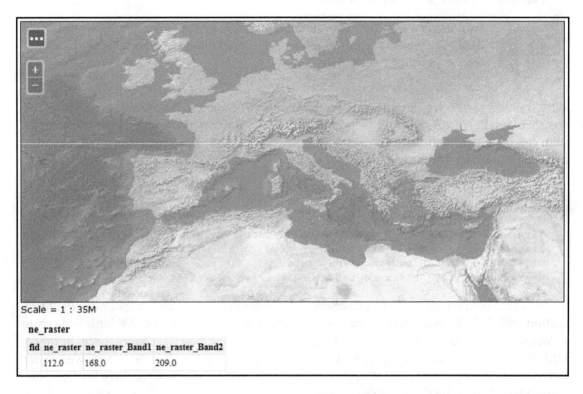

Scale = 1 : 35M

ne_raster

fid	ne_raster	ne_raster_Band1	ne_raster_Band2
112.0	168.0	209.0	

If you happen to preview the requests the browser is issuing, you may find that the URLs are similar to the following one that pulls an image for the Iberian peninsula:

```
http://localhost:8080/geoserver/mastering_postgis/wms?SERVICE=WMS&VERSI
ON=1.1.1&REQUEST=GetMap&FORMAT=image%2Fjpeg&TRANSPARENT=true&tiled=true
&STYLES&LAYERS=mastering_postgis%3Ane_raster&tilesOrigin=-179.983333333
3%2C-90.0166666658&WIDTH=256&HEIGHT=256&SRS=EPSG%3A4326&BBOX=-11.25%2C3
3.75%2C0%2C45
```

There is a GeoServer plugin meant to simplify all the work we have done so far. It is called PgRaster and you can obtain more information on it here: `http://geoserver.readthedocs.io/en/latest/community/pgras ter/pgraster.html`;

Older versions of the PGRaster plugin can be found here: `https://repo.b oundlessgeo.com/release/org/geoserver/community/gs-pgraster/`.

Outputting vector data as WFS services

Creating GeoServer layers to be outputted via WFS is exactly the same as for WMS, so we have already configured one layer. As an exercise, please configure a layer for the second table we imported, `ne_reefs`. When ready, it should look like this one when you preview it (note I have changed the color to red):

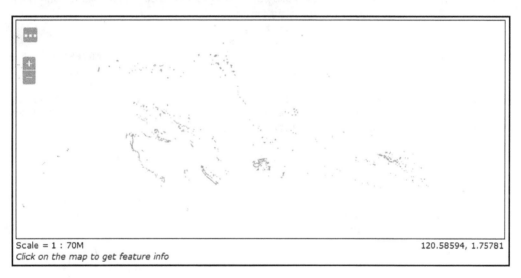

When you change the preview output to WFS, you will not see an image, but rather the output of a GetFeature request. By default, GeoServer limits output to 50 features; I have changed the URL to only return the first one though:

```
http://10.0.0.19:8080/geoserver/mastering_postgis/ows?service=WFS&versi
on=1.0.0&request=GetFeature&typeName=mastering_postgis:ne_coastline&max
Features=1&outputFormat=gml3
```

Thanks to that, we can paste the following XML:

```
<wfs:FeatureCollection numberOfFeatures="0"
timeStamp="2016-11-18T18:49:56.136Z"
xsi:schemaLocation="http://www.opengis.net/wfs
http://10.0.0.19:8080/geoserver/schemas/wfs/1.1.0/wfs.xsd mastering_postgis
http://10.0.0.19:8080/geoserver/mastering_postgis/wfs?service=WFS&version=1
.0.0&request=DescribeFeatureType&typeName=mastering_postgis%3Ane_coastline"
>
    <gml:featureMembers>
        <mastering_postgis:ne_coastline  gml:id="ne_coastline.1">
            <mastering_postgis:gid>1</mastering_postgis:gid>
<mastering_postgis:scalerank>6.00000000000</mastering_postgis:scalerank>
<mastering_postgis:featurecla>Coastline</mastering_postgis:featurecla>
            <mastering_postgis:geom>
                <gml:MultiLineString
srsName="http://www.opengis.net/gml/srs/epsg.xml#4326" srsDimension="2">
                    <gml:lineStringMember>
                        <gml:LineString>
                            <gml:posList>
                            -71.31509355 41.58079662 -71.30467689
41.60423412 -71.3033748 41.62636953 -71.31769772 41.63287995 -71.33071855
41.64720287 -71.33592689 41.65761953 -71.34373939 41.67064037 -71.35545814
41.6602237 -71.35936439 41.64199453 -71.35676022 41.63157787 -71.33592689
41.62636953 -71.33071855 41.61074453 -71.33722897 41.59251537 -71.33722897
41.5820987 -71.33071855 41.57558828 -71.31509355 41.58079662
                            </gml:posList>
                        </gml:LineString>
                    </gml:lineStringMember>
                </gml:MultiLineString>
            </mastering_postgis:geom>
        </mastering_postgis:ne_coastline>
    </gml:featureMembers>
</wfs:FeatureCollection>
```

Making use of PgRaster in a simple WMS GetMap handler

As mentioned previously, PgRaster support in GeoServer is not there yet. This is good, because we can learn how to consume it ourselves!

Let's import some data first:

```
raster2pgsql -s 4326 -C -l 2,4,6,8,10,12,14 -I -F -t 256x256
NE2_HR_LC_SR_W_DR.tif webgis.ne_raster | psql -h localhost -p 5434 -U
postgres -d mastering_postgis
```

In Chapter 5, *Exporting Spatial Data*, we used PostgreSQL's large object support to export the raster from the database. We will now build on what we achieved there, so we can come up with a simple raster extractor query for our WMS handler. The interesting bit is the query we used for assembling the tiles of the imported raster back into one raster:

```
select
    ST_Union(rast) as rast
from
    data_import.gray_50m_partial
where
    filename = 'gray_50m_partial_bl.tif'
```

Our slightly extended query looks like this:

```
select
    --3. union our empty canvas with the extracted raster
    ST_Union(rast) as rast
from (
    --1. empty raster based on the passed bounds and raster settings of
    the raster data is extracted from;
    --this is our 'canvas' we will paint the extracted raster on.
    --this lets us always output a raster that extends to the requested
    bounds
    select ST_AsRaster(ST_MakeEnvelope(14,85,24,95,4326), (select rast
    from data_import.gray_50m_partial limit 1)) as rast

    --2. extract the tiles of the raster that interset with bounds of out
    request and clip them to the requested bound
    union all select
        ST_Clip(
            ST_Union(rast),
            ST_MakeEnvelope(14,85,24,95,4326)
        ) as rast
    from
        data_import.gray_50m_partial

    where
        ST_Intersects(rast, ST_MakeEnvelope(14,85,24,95,4326))

) as preselect
```

As a matter of fact, we do not do much more. Basically, what happens here is:

- Generation of an empty raster that has the extent of the requested bounds and the parameters of the source raster we read from
- Extraction of the source tiles that intersect with the requested bounds; tiles are further cropped with the very same bounds
- Data assembly - we paint the natural earth raster on top of the empty canvas

At this stage, we have a raster we can output. To do so, we'll need a simple HTTP handler that can deal with our WMS requests. This will be a simplistic handler that is supposed to present the idea rather than be bullet-proof, production-ready code. Our WMS handler will only support a GetMap request.

Let's start with disassembling the WMS request itself into separate parameters, so it is clear what we are about to deal with. You may remember an example of a WMS request presented a few pages back - basically its query string will be similar to the following:
`?SERVICE=WMS&VERSION=1.1.1&REQUEST=GetMap&FORMAT=image%2Fjpeg&TRANSPARE NT=true&STYLES&LAYERS=mastering_postgis%3Ane_coastline&&WIDTH=256&HEIGH T=256&SRS=EPSG%3A4326&BBOX=-11.25%2C33.75%2C0%2C45.`

It will be a bit easier when we look at the parameters one by one:

- `SERVICE=WMS`: Service type
- `VERSION=1.1.1`: Version of the WMS service
- `REQUEST=GetMap`: Request type
- `FORMAT=image/jpeg`: Output format
- `TRANSPARENT=true`: Whether or not the background should be transparent; important when outputting vector data, but also when there are voids in the raster data
- `STYLES`: Styles to be applied to the requested data
- `LAYERS=mastering_postgis:ne_coastline`: Layers to be extracted
- `WIDTH=256`: Width of the image
- `HEIGHT=256`: Height of the image
- `SRS=EPSG:4326`: Coordinate system of the request
- `BBOX=-11.25,33.75,0,45`: Bounding box in the form of `minx,miny,maxx,maxy`

The bounding box described previously is specific to version < 1.3.0 of the WMS specification. Starting with version 1.3.0, the order of the bounding box coordinates depends on the order of coordinates defined by the SRS itself. Our WMS handler will only support v 1.1.1 of the specs.

Now, as we fully understand the WMS request parameters, we can move on to coding a simple handler in Node.js. We will build upon a web `Hello World` example that we wrote in a previous chapter:

 The code for this example can be found in the `code/wms` directory.

```
const http = require('http');
const url = require('url');

const server = http.createServer((req, res) => {

    console.warn('Processing WMS request...', req.url);

    var params = url.parse(req.url, true).query; // true to get query
    as object
    //fix param casing; url param names should not be case sensitive!...
    let pLowerCase = p.toLowerCase();
        if(p !== pLowerCase){
            params[pLowerCase] = params[p];
            delete params[p];
        }
    }

    //validate the request
    if(vaidateRequest(res, params)){
        processRequest(res, params);
    }
});

const port  = 8081; //another port so we can have it working with geoserver
server.listen(port,  () => {
    console.warn('WMS Server listening on http://localhost:%s', port);
});
```

As mentioned, our handler is going to be a simple one; therefore we will hardcode some logic that otherwise should be made dynamic, and thanks to that we'll keep the example clear.

Let's perform a simple checkup on the request parameter first so we can ensure that the submitted request is valid. According to the OGC specification, WMS should output exceptions in a specified form driven by the exception format parameter. In this case though, we will simply output 400, as handling exceptions the way that complies to the specs is not our task at this time. Our validation handler will therefore look like this:

```
/**
 * validates the WMS request; returns true if the request is valid and
false otherwise. if request is nod valid response writes 400 and closes
 */
const validateRequest = (res, params) => {
    var valid = true;
    try {
        for(var validator of validationRules){
            validator(params);
        }
    }
    catch(e){
        valid = false;
        handleError(res, e);
    }
    return valid;
}

/**
 * handles exception response
 */
const handleError = (res, msg) => {
    res.statusCode = 400;
    res.end(msg);
};
```

We'll need some validation rules too:

```
const validationRules = [
    (params) => {validateParamPresence(params, 'service')},
    (params) => {if(params.service !== 'WMS'){throw 'This service only
    supports WMS'}},
    (params) => {validateParamPresence(params, 'version')},
    (params) => {if(params.version !== '1.1.1'){throw 'The only supported
version
    is 1.1.1';}}, (...)
];
```

Do review the source code, as there are more validation rules applied to a request before it can be processed.

Once our request is validated, we should be able to render a map image safely (without errors). In order to do so, we have to talk to our database and for this, we need the pg module first:

```
npm install --pg save
```

Let's extract the params we need to process the request:

```
/**
 * generates wms output based on the params. params should be validated
prior to calling this method
 */
const processRequest = (res, params) => {

    //prepare some params first
    let w = parseInt(params.width);
    let h = parseInt(params.height);
    let bb = params.bbox.split(',');
    let minX = parseFloat(bb[0]);
    let minY = parseFloat(bb[1]);
    let maxX = parseFloat(bb[2]);
    let maxY = parseFloat(bb[3]);
    let format = getGdalFormat(params.format);

    //get table name based on tile resolution expressed in map units
    let tableName = getTableName(Math.abs(maxX - minX) / w);
}
```

The database connection skeleton is not very complex and actually it is pretty much what we have seen before:

```
//init client with the appropriate conn details
const client = new pg.Client({
    host: 'localhost',
    port: 5434,
    user: 'postgres',
    password: 'postgres',
    database: 'mastering_postgis'
});

//connect to the database
client.connect(function(err){
```

```
    if(err){
        handleError(err);
        return;
    }

    let query = `TODO`;

    client.query(query, function(err, result){
        client.end();

        if(err){
            handleError(err);
            return;
        }

        //handle response
        res.statusCode = 200;
        res.setHeader('content-type', params.format);
        res.end(result.rows[0].rast);
    });
});
```

The last part we have left is the actual query. We have experimented with it a bit, so it is now time to make it dynamic:

```
let query = `
select
    --3. union our empty canvas with the extracted raster and resize it to
the requested tile size
    ST_AsGDALRaster(
        ST_Resample(
            ST_Union(rast),
            $1::integer,
            $2::integer,
            NULL,NULL,0,0,'Cubic',0.125
        ),
        $3
    )as rast
from (
    --1. empty raster based on the passed bounds and raster settings of the
raster data is extracted from;
    --this is our 'canvas' we will paint the extracted raster on.
    --this lets us always output a raster that extends to the requested
bounds
    select ST_AsRaster(ST_MakeEnvelope($4,$5,$6,$7,4326), (select rast from
webgis.${tableName} limit 1)) as rast

    --2. extract the tiles of the raster that interset with bounds of out
```

```
request and clip them to the requested bound
    union all select
        ST_Clip(
            ST_Union(rast),
            ST_MakeEnvelope($4,$5,$6,$7,4326)
        )as rast
    from
        webgis.${tableName}
    where
        ST_Intersects(rast, ST_MakeEnvelope($4,$5,$6,$7,4326))
) as preselect
;
```

As you can see, the preceding query is almost the same as the one we saw already; the main difference is its parameterization. Also, I have added a ST_Resample call with the Cubic resampling algorithm, so the resized images look smooth, and an ST_AsGDALRaster call so we get the binary data that we can pipe straight into the response.

At this stage, our WMS handler should be ready, so let's launch it via the node index.js command and paste the following URL into the browser's address bar:
http://localhost:8081/?SERVICE=WMS&VERSION=1.1.1&REQUEST=GetMap&FORMAT=image%2Fpng&TRANSPARENT=true&STYLES&LAYERS=ne_raster&&WIDTH=256&HEIGHT=256&SRS=EPSG%3A4326&BBOX=-11.25%2C33.75%2C0%2C45.

The expected output is the Iberian Peninsula cut out of an o_2_ne_raster table:

Congratulations! You have just created your very own Geo-Server capable of serving WMS GetMap requests. Obviously, our implementation is quite limited and should be considered a rather basic one--our point though was to consume PgRaster, not to create a full- blown service.

Consuming WMS

Having exposed our WMS services, we can now consume them in a web application. The two most popular web mapping libraries are OpenLayers and Leaflet. We will have a closer look at both of them.

It is advised to review the documentation of both libraries. It can be found at `https://openlayers.org/` and `http://leafletjs.com/`.

Consuming WMS in ol3

Our first example renders a Window with an OpenLayers map inside. We add three layers to the map - all of them exposed via GeoServer. We need to create a map with the following code:

```
/**
 * Create map
 * @param mapContainerId
 */
createMap: function(mapContainerId) {

    var proj = ol.proj.get('EPSG:4326');

    this.map = new ol.Map({
        layers: this.createLayers(),
        target: mapContainerId,
        controls: ol.control.defaults({
            attributionOptions: {
                collapsible: false
            }
        }).extend([
            new ol.control.ScaleLine(),
            new ol.control.MousePosition({
                projection: proj,
                coordinateFormat: function(coords) {
                    var output = '';
                    if (coords) {
                        output = coords[0].toFixed(5) + ' : ' +
```

```
                              coords[1].toFixed(5);
                   }
                   return output;
               }
           })
       ]),

       view: new ol.View({
           projection: proj,
           extent: proj.getExtent(),

           center: [155, -15], //Australian Great Coral Reef
           zoom: 5
       })
   });
},
```

The preceding code creates a map with some simple controls and the layers specified by the following method:

```
/**
 * creates layers for the map
 * @returns {[*]}
 */
createLayers: function() {
    var proj = ol.proj.get('EPSG:4326');
    return [
        new ol.layer.Tile({
            source: new ol.source.TileWMS({
                url: 'http://localhost:8080/geoserver/wms',
                params: {
                    'LAYERS': 'mastering_postgis:ne_coastline,
                    mastering_postgis: ne_reefs ',
                    'VERSION': '1.1.1'
                },
                projection: proj,
                extent: proj.getExtent(),
                attributions: [
                    new ol.Attribution({
                        html: 'Mastering PostGIS - GeoServer vector'
                    })
                ]
            })
        })
    ];
}
```

We now need to navigate to the example's folder, run `sencha app watch`, and navigate to `http://localhost:1841/apps/02_ol3_wms/`. You should see a simple map of the Australian Great Barrier Reef:

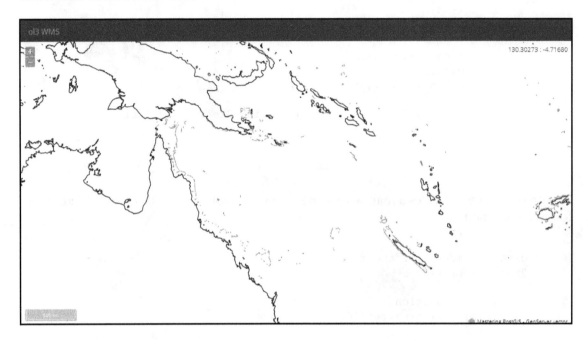

Let's add a bit of life to our map and also display the raster that we loaded using the `ImageMosaicJDBC` plugin:

```
new ol.layer.Tile({
    source: new ol.source.TileWMS({
        url: 'http://localhost:8080/geoserver/wms',
        params: {
            'LAYERS': 'mastering_postgis:ne_raster',
            'VERSION': '1.1.1'
        },
        projection: proj,
        extent: proj.getExtent(),
        attributions: [
            new ol.Attribution({
                html: 'Mastering PostGIS - GeoServer raster'
            })
        ]
    })
}),
```

The preceding code is just a layer definition that pulls data for our raster layer. When we add it above the WMS layer reading vector data, we should get the following output:

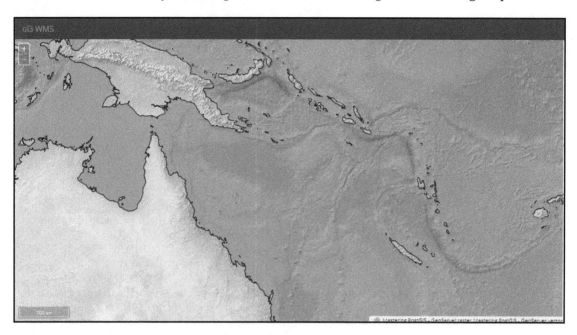

Consuming WMS in Leaflet

Map declaration in Leaflet is very similar to what we did with OpenLayers:

```
/**
 * Create map
 * @param mapContainerId
 */
createMap: function(mapContainerId){
    this.map = new L.Map(mapContainerId, {
        crs: L.CRS.EPSG4326,
        layers: this.createLayers()
    }).setView([-15,155], 4);

    L.control.scale().addTo(this.map);
}
```

The preceding code creates a Leaflet Map instance and renders it into the specified container. Layers are declared in another method:

```
/**
 * creates layers for the map
 * @returns {[*]}
 */
createLayers: function(){
    return [
        L.tileLayer.wms('http://localhost:8081', {
            layers: 'ne_raster',
            version: '1.1.1',
            format: 'image/png',
            transparent: true,
            maxZoom: 8,
            minZoom: 0,
            continuousWorld: true,
            attribution: 'Mastering PostGIS - NodeJs WMS handler reading
            pgraster'
        }),
        L.tileLayer.wms('http://localhost:8080/geoserver/wms?', {
            layers:
            'mastering_postgis:ne_coastline,mastering_postgis:ne_reefs',
            version: '1.1.1',
            format: 'image/png',
            transparent: true,
            maxZoom: 8,
            minZoom: 0,
            continuousWorld: true,
            attribution: 'Mastering PostGIS - GeoServer vector'
        })
    ];
}
```

Once again, we add two WMS layers to a map: first a layer reading raster data, and next a layer rendering vector data. This time though we made a little change: our raster data WMS endpoint is now the Node.js- powered service.

We now need to navigate to the example's folder, run `sencha app watch`, and navigate to `http://localhost:1841/apps/03_leaflet_wms/`.

As expected, the image is very similar to the one we saw already rendered by ol3. Basically, the only noticeable difference is the map controls and the panel title:

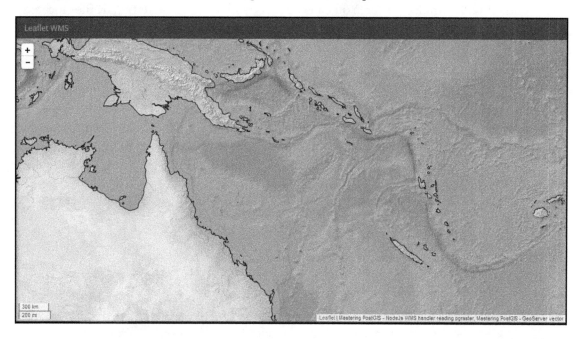

Enabling CORS in Jetty

Because our GeoServer is hosted at a different origin than our web application (a different port is enough to make a domain be considered by a browser to be a different origin; an origin is a combination of protocol, host, and port), we will not be able to perform AJAX requests straight away as the browser will refuse to retrieve the data from such a location. This is due to the same-origin policy that is meant to prevent scripts from untrusted sources gaining access to the DOM of a page.

CORS (Cross Origin Resource Sharing) is a standard mechanism for cross origin communication between browsers and servers. The CORS specification defines a set of headers that are used to communicate which operations are allowed. Thanks to that, it is possible to expose APIs that can be consumed by web clients located in different domains than the API itself.

In order to enable our web apps to send AJAX requests to our remote GeoServer, we need to enable CORS in our Jetty server. In order to do so, first we need to check the version of Jetty bundled with our GeoServer. You can check it by looking at the Jetty JAR files located in the `geoserver/lib` directory. In my case, it is Jetty `9.2.13.v20150730`. Next, we need to obtain an appropriate servlets file from `http://repo1.maven.org/maven2/org/eclipse/jetty/jetty-servlets/`. In my case, it was `http://repo1.maven.org/maven2/org/eclipse/jetty/jetty-servlets/9.2.13.v20150730/jetty-servlets-9.2.13.v20150730.jar`. Once downloaded, the servlets JAR file should be put in `webapps/geoserver/WEB-INF/lib`. The last step is to modify `webapps/geoserver/WEB-INF/web.xml` and add the following XML (I have put the mine just after the context-param declarations and before the first filter declaration):

```
<filter>
    <filter-name>cross-origin</filter-name>
    <filter-class>org.eclipse.jetty.servlets.CrossOriginFilter</filter-
class>
</filter>
<filter-mapping>
    <filter-name>cross-origin</filter-name>
    <url-pattern>/*</url-pattern>
</filter-mapping>
```

If you happen to have a different version of Jetty, or you have deployed your GeoServer in Tomcat, please refer to: `http://enable-cors.org/` for detailed instructions on how to enable CORS.

At this stage, our server should be CORS-enabled and our cross origin AJAX examples should work as expected without having to use a pass-through proxy in order to connect to a service located in a different domain.

Consuming WFS in ol3

Leaflet does not read WFS, so our WFS example will be OpenLayers only. Consuming WFS means reading, retrieving, and rendering vector data on the client side. Such processes may be intensive on PC resources - we will therefore use a rather small dataset with the `ne_reefs` vector so we do not have to bother with performance issues when dealing with larger vector datasets.

There are different techniques for dealing with large datasets, such as the bounding box strategy, mixing WMS for smaller scales, displaying with WFS for larger scale display, and using vector tiles. Those are not specific to the data source as such, and therefore I will not elaborate on the subject.

Our web example will be very similar to the previous one with WMS only. The different part is the layers declaration:

```
createLayers: function(){
    var proj = ol.proj.get('EPSG:4326');
        format = new ol.format.WFS(),
        wfsVectorSource = new ol.source.Vector({
            projection: proj,
            loader: function(extent, resolution, projection) {
                Ext.Ajax.request({
                    cors: true,
                    url: 'http://localhost:8080/geoserver/wfs?
                    service=WFS&request=GetFeature&version=1.1.0&' +
                    'typename=mastering_postgis:ne_reefs&'+
                    'srsname=EPSG:4326&' +
                    'bbox=' + extent.join(',') + ',EPSG:4326'
                })
                .then(function(response){
                    //rad the features
                    var features =
                    format.readFeatures(response.responseText),
                        f = 0, flen = features.length;

                    //and make sure to swap the coords...
                    for(f; f < flen; f++){
                        features[f].getGeometry().applyTransform(function
                        (coords, coords2, dimension) {
                            var c = 0, clen = coords.length,
                                x,y;
                            for (c; c < clen; c += dimension) {
                                y = coords[c]; x = coords[c + 1];
                                coords[c] = x; coords[c + 1] = y;
                            }
                        });
                    }
                    wfsVectorSource.addFeatures(features);
                });
            },
            //this will make the map request features per tile boundary
            strategy: ol.loadingstrategy.tile(ol.tilegrid.createXYZ({
                extent: proj.getExtent(),
                maxZoom: 8
```

```
                    })),
                    attributions: [
                        new ol.Attribution({
                            html: 'Mastering PostGIS - GeoServer WFS'
                        })
                    ]
            });

        return [
            (...WMS Layer declaration    ),
            new ol.layer.Vector({
                source: wfsVectorSource,
                style: style: (...style declaration...)
            })
        ];
    }
```

The preceding code is a bit more complex than a simple WMS layer declaration, as we need to declare a format object that can read the GML returned by the WFS services, and also, since a vector loading strategy is used, we need a loader function for it. If this was not enough, we need to do some coordinate swapping as EPSG:4326 reverses the axis order.

We now need to navigate to the example's folder, run sencha app watch, and navigate to http://localhost:1841/apps/04_ol3_wfs/. You should see a similar output to the following:

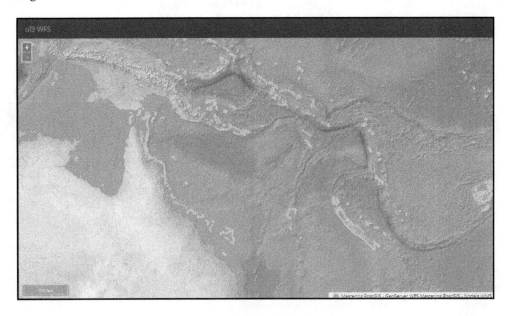

Outputting and consuming GeoJSON

GML output may be considered by many as not too user-friendly, as XML in general is supposed to be machine readable, but not necessarily human readable (although once one gets used to it - it is not that scary anymore). Luckily, servers such as GeoServer can output data in GeoJSON. This makes things way easier, not only from the perspective of the user looking at the output, but also when it comes to coding the layers that consume it.

This time we will plug in some earthquake data - we will need the data we imported in Chapter 1, *Importing Spatial Data*, but this time let's put it into the `webgis.earthquakes` table.

 If you happen to have removed the data from the `imported_data` schema, you will have to reimport the dataset. Otherwise, you can simply copy the data over with `select * into webgis.earthquakes from imported_data.earthquakes_subset_with_geom`.

Next, make sure to expose the data as a `mastering_postgis:earthquakes` layer and we are ready to roll.

 Even though we consume GeoJSON now, the service we request data from is still WFS.

Consuming GeoJSON in ol3

GeoJSON layer declaration in ol3 is way simpler than what we had to do with WFS. This time, BBOX strategy is used to load the data for whatever area is visible on the map:

```
createLayers: function(){

    var proj = ol.proj.get('EPSG:4326');

    return [
        (...WMS Layer declaration),
        new ol.layer.Vector({
            //projection: proj,
            source: new ol.source.Vector({
                format: new ol.format.GeoJSON(),
                url: function(extent) {
                    return 'http://localhost:8080/geoserver/wfs?service=WFS&' +
                    'version=1.1.0&request=GetFeature&typename=mastering
```

```
            _postgis:earthquakes&' +
            'outputFormat=application/json&srsname=EPSG:4326&' +
            'bbox=' + extent.join(',') + ',EPSG:4326';
        },
        strategy: ol.loadingstrategy.bbox,
        attributions: [
            new ol.Attribution({
                html: 'Mastering PostGIS - GeoServer GeoJSON'
            })
        ]
    }),
    style: (...style declaration...)
  })
];
}
```

We now need to navigate to the example's folder, run `sencha app watch`, and navigate to `http://localhost:1841/apps/05_ol3_geojson/`. You should see a similar output to the following:

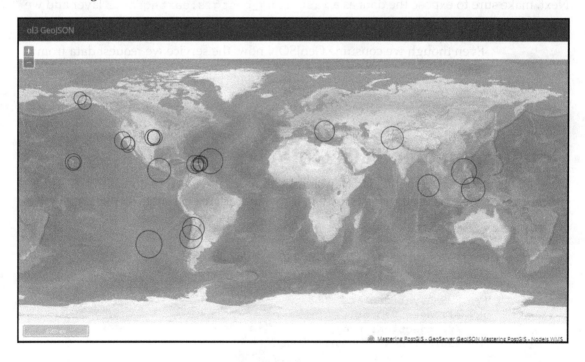

Consuming GeoJSON in Leaflet

The Leaflet GeoJSON layer does not retrieve the data automatically and therefore we need to obtain it ourselves. Once data is present, we can load it with the following code:

```
loadGeoJSON: function(map){
    Ext.Ajax.request({
        cors: true,
        url: 'http://10.0.0.19:8080/geoserver/wfs?service=WFS&' +
        'version=1.1.0&request=GetFeature&typename=
        mastering_postgis:earthquakes&' +
         'outputFormat=application/json&srsname=EPSG:4326&' +
         'bbox=-180,-90,180,90,EPSG:4326'
    }).then(function(response){
        L.geoJson(Ext.JSON.decode(response.responseText).features, {
            pointToLayer: function (feature, latlng) {
                return L.circleMarker(latlng, {
                    radius: feature.properties.mag * 5,
                    fillColor: '#993366',
                    weight: 1
                });
            }
        }).addTo(map);
    });
},
```

As you can see, it is way simpler than consuming GeoJSON in ol3. We now need to navigate to the example's folder, run `sencha app watch`, and navigate to `http://localhost:1841/apps/06_leaflet_geojson/`. You should see a similar output to the following:

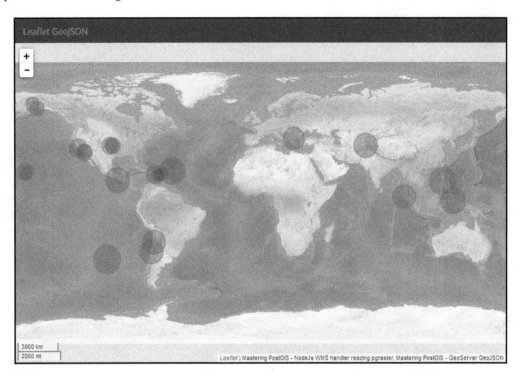

Outputting and consuming TopoJSON

Since we're consuming vectors in a browser, let's have a look at TopoJSON too. TopoJSON is an extension of GeoJSON--the main difference is it encodes geometries shared by multiple features only once, so it is possible to reduce the footprint of the returned data. It is a popular format when it comes to serving vector tiles too. We will not get into detail on how to set up our own tile server; instead we will make our PostGIS output the data for us.

There is a GeoServer plugin for exposing TopoJSON called vector tiles; some Node modules that can do the job are also available--just search npm for TopoJSON.

Let's get some data into the database first: download
`http://www.naturalearthdata.com/http//www.naturalearthdata.com/download/50m/cultural/ne_50m_admin_0_countries.zip` and load it into the `webgis.countries` table:

```
shp2pgsql -s 4326 ne_50m_admin_0_countries webgis.countries | psql -h
localhost -p 5434 -U postgres -d mastering_postgis
```

We will need to process the data before we can use the `topology.AsTopoJSON` function because it accepts topo geometry. So let's do the data preparation first:

```
--creates a new topology schema and registers it in the topology.topology
table
select topology.CreateTopology('topo_countries', 4326);

--a new table that will hold topogeom
create table webgis.countries_topo(git serial primary key, country
varchar);
--add topogeom col to a table and registers it as a layer in the
topology.layer table
select topology.AddTopoGeometryColumn('topo_countries', 'webgis',
'countries_topo', 'topo', 'MULTIPOLYGON');

--this will convert the original country geoms into topo geoms
insert into webgis.countries_topo (country, topo)
select
    name,
    topology.toTopoGeom(
        geom,
        'topo_countries',
        --third param is topology layer identifier; we obtain it
automatically based on our topology name
        (select layer_id from topology.layer where schema_name = 'webgis'
and table_name = 'countries_topo' limit 1),
        0.00001 --note: precision param. needed, so we avoid problems with
invlaid geoms
    )
from
    webgis.countries;
```

The preceding query simply prepares a new table with topo geometry based on the original geometry imported from a shapefile. Once our topo geometry is ready, we can move on to generating TopoJSON.

The following code is a simple adaptation of the example presented on the `AsTopoJSON` function documentation page (`http://postgis.net/docs/manual-dev/AsTopoJSON.html`). It adds a file export and does some cleanup too, so we do not leave the mess behind:

```
DROP TABLE IF EXISTS edgemap;
CREATE TEMP TABLE edgemap(arc_id serial, edge_id int unique);

DROP TABLE IF EXISTS topojson;
CREATE TEMP TABLE topojson(json_parts varchar);

INSERT INTO topojson

-- header
SELECT '{ "type": "Topology", "transform": { "scale": [1,1], "translate":
[0,0] }, "objects": {'

-- objects already stitched together
UNION ALL select array_to_string( array_agg(json_parts), E', ')
from  (SELECT '"' || country || '": ' || topology.AsTopoJSON(topo,
'edgemap') as json_parts
   FROM webgis.countries_topo) as json_parts;

-- arcs
WITH edges AS (
  SELECT m.arc_id, e.geom FROM edgemap m, topo_countries.edge e
  WHERE e.edge_id = m.edge_id
), points AS (
  SELECT arc_id, (st_dumppoints(geom)).* FROM edges
), compare AS (
  SELECT p2.arc_id,
         CASE WHEN p1.path IS NULL THEN p2.geom
             ELSE ST_Translate(p2.geom, -ST_X(p1.geom), -ST_Y(p1.geom))
         END AS geom
  FROM points p2 LEFT OUTER JOIN points p1
  ON ( p1.arc_id = p2.arc_id AND p2.path[1] = p1.path[1]+1 )
  ORDER BY arc_id, p2.path
), arcsdump AS (
  SELECT arc_id, (regexp_matches( ST_AsGeoJSON(geom), '\[.*\]'))[1] as t
  FROM compare
), arcs AS (
  SELECT arc_id, '[' || array_to_string(array_agg(t), ',') || ']' as a FROM
arcsdump
  GROUP BY arc_id
  ORDER BY arc_id
)
```

```
INSERT INTO topojson

SELECT '}, "arcs": [' UNION ALL
SELECT array_to_string(array_agg(a), E', ') from arcs

-- json footer part
UNION ALL SELECT ']}'::text;

--finally dump the topojson
COPY (SELECT array_to_string( array_agg(json_parts), E' ') FROM (SELECT
json_parts FROM topojson) AS json_parts)  TO 'f:\topojson.json';

--cleanup the temp stuff
DROP TABLE IF EXISTS edgemap;
DROP TABLE IF EXISTS topojson;
```

At this stage, our TopoJSON file should be ready, so the final thing is to display it on a map.

 You may actually want to test the output first - if so, simply upload the file to `http://geojson.io/`.

Consuming TopoJSON in ol3

So let's code an ol3 example. Once again, we are merely modifying a layer definition:

```
new ol.layer.Vector({
    //projection: proj,
    source: new ol.source.Vector({
        format: new ol.format.TopoJSON(),
        url: 'data/topojson.json',
        attributions: [
            new ol.Attribution({
                html: 'Mastering PostGIS - TopoJSON'
            })
        ]
    }),
    style: [
        new ol.style.Style({
            stroke: new ol.style.Stroke({
                color: 'rgba(0, 0, 0, 1)',
                width: 0.5
            })
        })
    ]
```

```
})
```

Because we need the JSON to be read by our application, make sure that the file is available under `data/topojson.json` in the root of our web example.

In order to preview the example, navigate to the example's folder, run `sencha app watch`, and navigate to `http://localhost:1841/apps/07_ol3_topojson/`. You should see a similar output to the following:

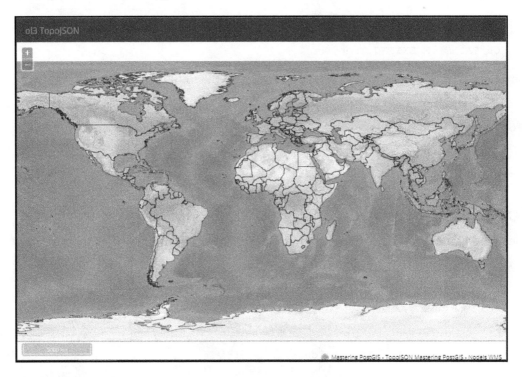

Consuming TopoJSON in Leaflet

By now, you should expect the Leaflet examples to be rather compact and easy to understand. You are going to be surprised though when you see that consuming a TopoJSON layer in Leaflet is a matter of one line:

```
omnivore.topojson('data/topojson.json').addTo(map);
```

In order to achieve this though, we need a MapBox Leaflet plugin called **omnivore**. You can get some more information on it here:

`https://www.mapbox.com/mapbox.js/example/v1.0.0/omnivore-topojson/.`

> Because we need the JSON to be read by our application, make sure that the file is available under `data/topojson.json` in the root of our web example.

The exact source code of the example is a bit longer, as there is a style declaration and I feed the TopoJSON to a GeoJSON layer, so the style can be applied.

In order to preview the example, navigate to the example's folder, run `sencha app watch`, and navigate to `http://localhost:1841/apps/08_leaflet_topojson/`. You should see a similar output to the following:

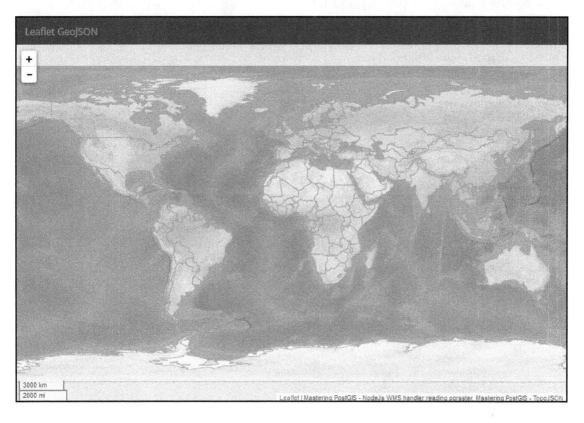

Implementing a simple CRUD application that demonstrates vector editing via web interfaces

Our final example in this chapter demonstrates some simple spatial CRUD functionality. CRUD stands for Create, Read, Update, Destroy, so to put it simply we'll edit some features on the map. We will do it using ol3.

WebGIS CRUD server in Node.js

In order to expose our `crud` API, we will need some storage for our features, so let's start with creating a table of the geometries first:

```
create table webgis.crud (id serial NOT NULL, geom geometry);
```

Once this is ready, we will need a simple web server to handle our CRUD operations. Let's take care of that.

First, we need to put some packages in place:

```
npm install express --save
npm install body-parser --save
npm install pg --save
```

Once the appropriate packages are installed, let's create our server:

```
const pg = require("pg");
const express = require('express');
const app = express();

const bodyParser = require('body-parser');

// configure app to use bodyParser() so we can get data from POST & PUT
app.use(bodyParser.urlencoded({ extended: true }));
app.use(bodyParser.json());

const dbCredentials = {
    host: 'localhost',
    port: 5434,
    user: 'postgres',
    password: 'postgres',
    database: 'mastering_postgis'
};
```

```
//express server
const server = app.listen(8082, () => {
  console.log(`WebGIS crud server listening at
http://${server.address().address}:${server.address().port}`);
});

/**
 * enable CORS
 */
app.use((req, res, next) => {
  res.header("Access-Control-Allow-Origin", "*");
  res.header("Access-Control-Allow-Headers", "Origin, X-Requested-With,
Content-Type, Accept");
res.header("Access-Control-Allow-Methods", "POST, GET, PUT, DELETE,
OPTIONS");
  next();
});

//prefix all the routes with 'webgisapi'
let router = express.Router();
app.use('/webgisapi', router);

/**
 * errorneus response sending helper
 */
const sendErrorResponse = (res, msg) => {
    res.statusCode = 500;
    res.end(msg);
}
```

This is a fully operational Express server, although in order to make it work for us we will need some logic to handle our requests, so let's create it.

Read method:

```
router.route('/features').get((req, res) => {

    //init client with the appropriate conn details
    let client = new pg.Client(dbCredentials);

    client.connect(function(err){
        if(err){
            sendErrorResponse(res, 'Error connecting to the database: ' +
            err.message);
            return;
        }

        //once connected we can now interact with a db
```

```
        client.query('SELECT id, ST_AsText(geom) as wkt FROM webgis.crud;',
        (err, result) => {
            //close the connection when done
            client.end();

            if(err){
                sendErrorResponse(res, 'Error reading features: ' +
                err.message);
                return;
            }

             if(result.rows.length === 0){
                res.statusCode = 404;
            }
            else {
                res.statusCode = 200;
            }

            res.json(result.rows);
        });
    });
});
```

When you visit `http://localhost:8082/webgisapi/features` now, you should get
the content of our database, although it is going to be just an empty array. Also, the
response code for no results will be 404 - not found.

Create method:

```
router.route('/features').post((req, res) => {
    //init client with the appropriate conn details
    let client = new pg.Client(dbCredentials);

    client.connect((err) => {
        if(err){
            sendErrorResponse(res, 'Error connecting to the database: ' +
            err.message);
            return;
        }

        //extract wkt off the request
        let wkt = req.body.wkt;

        //once connected we can now interact with a db
        client.query('INSERT INTO webgis.crud (geom) values
        (ST_GeomFromText($1, 4326)) RETURNING id;',[wkt], (err, result) =>
{
            //close the connection when done
```

```
        client.end();

        if(err){
            sendErrorResponse(res, 'Error reading features: ' +
            err.message);
            return;
        }

        res.statusCode = 200;
        res.json({id: result.rows[0].id, wkt: wkt});
    });
  });
});
```

The update method is very similar to the create code that we just wrote. Let's have a look at it now:

```
router.route('/features/:feature_id').put((req, res) => {
    //init client with the appropriate conn details
    let client = new pg.Client(dbCredentials);

    client.connect((err) => {
        if(err){
            sendErrorResponse(res, 'Error connecting to the database: ' +
            err.message);
            return;
        }

        //extract wkt off the request and id off the params
        let wkt = req.body.wkt;
        let id = req.params.feature_id;

        //once connected we can now interact with a db
        client.query('UPDATE webgis.crud set geom = ST_GeomFromText($1,
        4326) where id = $2;',[wkt, id], (err, result) => {
            //close the connection when done
            client.end();

            if(err){
                sendErrorResponse(res, 'Error reading features: ' +
                err.message);
                return;
            }

            res.statusCode = 200;
            res.json({id: id, wkt: wkt});
        });
    });
```

```
    });
```

As you may expect already, the Destroy method is also going to be very simple:

```
router.route('/features/:feature_id').delete((req, res) => {
    //init client with the appropriate conn details
    let client = new pg.Client(dbCredentials);

    client.connect((err) => {
        if(err){
            sendErrorResponse(res, 'Error connecting to the database: ' +
            err.message);
            return;
        }

        //extract id off the params
        let id = req.params.feature_id;

        //once connected we can now interact with a db
        client.query('DELETE FROM webgis.crud where id = $1;',[id], (err,
        result) => {
            //close the connection when done
            client.end();

            if(err){
                sendErrorResponse(res, 'Error reading features: ' +
                err.message);
                return;
            }

            res.statusCode = 200;
            res.json({id: id, wkt: wkt});
        });
    });
});
```

We have one API method to do before we can consider it complete. This time, we are after buffering geometry passed from the client:

```
router.route('/features/buffers').post((req, res) => {
    //init client with the appropriate conn details
    let client = new pg.Client(dbCredentials);

    client.connect((err) => {
        if(err){
            sendErrorResponse(res, 'Error connecting to the database: ' +
            err.message);
            return;
        }

        //extract wkt off the request
        let wkt = req.body.wkt;
        let buffer = req.body.buffer;

        //once connected we can now interact with a db
        client.query('SELECT ST_AsText(ST_Buffer(ST_GeomFromText($1, 4326),
        $2)) as buffer;',[wkt, buffer], (err, result) => {
            //close the connection when done
            client.end();

            if(err){
                sendErrorResponse(res, 'Error reading features: ' +
                err.message);
                return;
            }

            res.statusCode = 200;
            res.end(result.rows[0].buffer);
        });
    });
});
```

As you can see, implementing a barebones CRUD backend for our spatial database is not very difficult. Obviously, we do not handle permissions; there is no authentication, security, and so on. This should be implemented before deploying the application for others to use. Luckily, we can remain with our toys in a sandbox.

WebGIS CRUD client

In order to preview the example, navigate to the example's folder, run `sencha app watch`, and then navigate to `http://localhost:1841/apps/09_crud/`. You should see a similar output to the following:

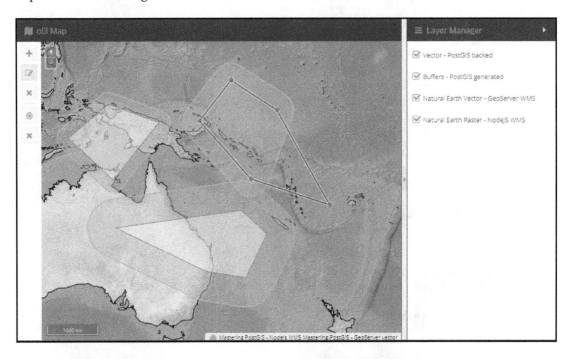

Our web editing functionality is implemented in ol3. This is because ol3 provides some ready-to-use, out-of-the box classes that can be used to quickly implement the WebGIS experience. Web editing is obviously possible in Leaflet too, but requires obtaining some plugins that extend Leaflet appropriately.

If you happen to like Leaflet more than ol3, nothing is lost. The code accompanying this example has been separated in a way; the db data exchange part is generic and can be used by code other than based on ol3. Also, I have created a Leaflet app stub so you have something to start with.

The client application can be split into functional parts that are going to be discussed next:

- Layer manager
- CRUD tools
- Analytical toolset

Because handling the UI interactions does not really involve the database, and rather focuses on gathering the user input and passing it on to the API for processing, the client-side code will be limited, so you can get an idea of what is going on, without having to review the full logic. It is advised that you review the accompanying source code in order to get the insight.

Layer manager

We already saw the layer declarations for some different types of underlying data. A WebGIS application is usually tasked with working with multiple layers, so having a layer manager of some sort seems to be a reasonable requirement. We're not going to be getting into details about how this has been achieved as it is simply down to showing a couple of checkboxes that turn layers on/off. Also, the source code is available so you can check it out easily. Our manager is a static one - it does not allow layer reordering. After all, our application is meant to be rather simple.

Drawing tools

OpenLayers3 has some controls that provide functionality such as drawing features, modifying features, or selecting them. Our examples use three such controls in different combinations: ol.interaction.Draw, ol.interaction.Select, and ol.interaction.Modify. The names are self-explanatory, so let's see how the feature add' functionality is implemented:

```
onBtnAddToggle: function(btn, state){
    this.clearMapInteractions();
    if(!state){
        return;
    }

    this.currentInteractions = {
        draw: new ol.interaction.Draw({
            type: 'Polygon',
            source: this.vectorLayer.getSource()
        })
    };
    this.currentInteractions.draw.on('drawend', this.onDrawEnd, this);

    this.map.addInteraction(this.currentInteractions.draw);
}
```

The logic starts with a call to a function that takes care of disabling any controls that may have been active before. Then, if the **Add feature** button has been depressed, a Draw interaction is created and added to the map. A `drawend` event is wired up to the control so we can react whenever a feature has been added:

```
onDrawEnd: function(e){
    this.saveFeature({
        wkt: this.getWktFormat().writeGeometry(e.feature.getGeometry())
    });
}
```

When a feature has been created, it is redirected to a proxy class that is responsible for handling the actual save procedure.

Editing a feature uses two controls - select interaction and modify interaction - so the user can first select a feature, and then edit it:

```
onBtnEditToggle: function(btn, state){
    this.clearMapInteractions();

    if(!state){
        return;
    }

    var select = new ol.interaction.Select({
        layers: [this.vectorLayer],
        style: this.getEditSelectionStyle()
    });
    this.currentInteractions = {
        select: select,
        modify: new ol.interaction.Modify({
            features: select.getFeatures(),
            style: this.getEditStyle()
        })
    };

    this.currentInteractions.select.on('select', this.onModifyStartEnd,
    this);

    this.map.addInteraction(this.currentInteractions.select);
    this.map.addInteraction(this.currentInteractions.modify);
}
```

As you can see, the editing functionality setup is quite easy too. This time we instantiate select and modify interactions, but listen only to the `select` event. This is because it is assumed that an edit starts once a feature is selected and it ends once the user decides to deselect the feature:

```
onModifyStartEnd: function(e){
    console.warn('[ol3] - modify start/end', e);

    if(e.selected.length > 0){
        //this is a select so just a start of edit
        //simply store a wkt on a feature so can compare it later and
        decide if an edit should happen
        e.selected[0].tempWkt =
        this.getWktFormat().writeGeometry(e.selected[0].getGeometry());
        return;
    }

    var f = e.deselected[0],
        modifiedWkt = this.getWktFormat().writeGeometry(f.getGeometry());

    if(f.tempWkt === modifiedWkt){
        console.warn('[ol3] - modify end - feature unchanged');
        return;
    }

    console.warn('[ol3] - modify end - saving feature...');

    this.saveFeature({
        id: f.get('id'),
        wkt: modifiedWkt
    });
}
```

We can arrive in the `select` handler in two scenarios: when a user either selects or deselects a feature. On select, a snapshot of the feature geometry is taken so it can be compared with another snapshot taken when handling deselect. If the geometry has changed, the modified object is delegated to our proxy class for saving:

```
saveFeature: function(f){

    console.warn('[CRUD PROXY] - saving...', f);

    Ext.Ajax.request({
        cors: true,
        url: isNaN(f.id) ? apiEndPoint : (apiEndPoint + f.id),
        method: isNaN(f.id) ? 'POST' : 'PUT',
        success: Ext.bind(this.onSaveSuccess, this),
```

```
        failure: Ext.bind(this.onSaveFailure, this),
        params: f
    });
}
```

Depending on ID presence, we send out either a `POST` request (creating a feature) or a `PUT` request (modifying a feature). For the sake of simplicity, once an operation succeeds, the map data is fully reloaded. This is far from being optimal, but lets us avoid syncing feature state.

The delete control is very similar to what we saw when editing, it doesn't use the modify interaction though. Whenever a user selects a feature, they are asked if they wish to remove a feature and if so, a `DELETE` request is made to the server. When finished, our feature set is re-read from the database.

Analysis tools - buffering

There are many analytics tools available in PostGIS, so buffering may seem a bit dull. On the other hand, it is a common example and it is simple enough to let us focus on how to use it in web realms. Thanks to that, we can quickly demonstrate a simple logic that can later be scaled into some more sophisticated tools.

Our buffer functionality is almost the same as the delete functionality. It uses the same type of interaction - select interaction - but instead of asking the user for a confirmation, it prompts the user to input a buffer size in degrees. Once the data is collected, a POST request is sent to the server, which replies with a WKT encoded buffer geometry.

Summary

I hope that this chapter has demystified WebGIS a bit. Having a powerful tool watch our backs (PostGIS of course!) is a good starter to writing our very own web-based GIS applications.

We have managed to touch upon many different tools in this chapter; we have also written our own WMS server and REST API, and consumed both along with GeoServer services. This is quite a lot, even though our examples are simple and focused.

When outputting TopoJSON we have touched topo geometry; if this seemed a bit unfamiliar stay tuned - the next chapter is all about PostGIS topology.

8
PostGIS Topology

PostGIS originally implemented only the Simple Features model for storing vector data. In this model, every feature is a distinct entity, and any topological relationships between them aren't explicitly stored in a database. For some use cases, this data model is not a good fit. Two notable examples include boundary data and network data. In boundary data, shared lines between adjacent features aren't really shared: the features are stored as separate polygons and the same set of vertices is stored twice. In network data, the database doesn't store any information about shared nodes. This makes spatial analysis harder (relationships have to be checked on a vertex-by-vertex basis, which is time-consuming). Even worse, editing or simplification of data can lead to inconsistencies, including gaps between contiguous polygons, overlapping polygons, or disconnected lines.

To address these issues, a topology model for PostGIS has been introduced. In a topological model, spatial relationships between connected features are explicitly modeled. The topology extension has been around since version 1.1, but in PostGIS 2.0 it became a first-class citizen, fully functional and included in a standard installation. In this chapter, we will discuss the features and use of the PostGIS topology extension for storing and processing geospatial data.

The conceptual model

Before we start working with topology, we need to discuss the key concepts behind the PostGIS topological data model. PostGIS topology implementation is based on *ISO standard 13249 - Information technology - Database languages - SQL multimedia and application packages - Part 3: Spatial*. The standard name is often abbreviated as **ISO SQL/MM**. This standard defines two data models for topology: `TopoGeometry` and `TopoNetwork`; PostGIS implements only the former.

In this model, there are three kinds of elements/primitives used to compose geospatial features: nodes, edges, and faces.

Nodes are point features. They can exist on their own (isolated nodes) or serve as connection points for linear features (the edges). For example, this is a Czech-Polish-Slovak tripoint near Jaworzynka village, which contains a node and three edges:

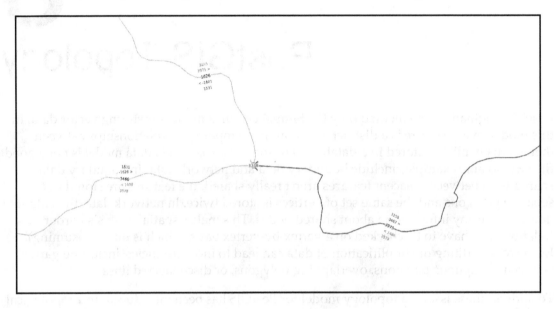

An example of topological data: a node and three edges

In topology, only one node can appear for a specified coordinate pair - if there are more nodes in the same place, the topology is invalid.

If you are familiar with the OpenStreetMap data model, which is also topological, note that only intersections - and not all vertices - are considered nodes in PostGIS. If you are familiar with network data models, note that node cardinality (that is, how many edges are connected to a node) is not stored in a database, but it can be retrieved using a non-spatial SQL query.

Edges are linear features. They must start precisely at one node and end at another node; both nodes' IDs are stored in a database. If there is a mismatch between the node ID and its position, or if an edge crosses a node without ending, the topology is considered invalid.

Faces are used to represent polygons. Their precise geometry is not stored, only the polygon's bounding boxes, called **MBR** from **Minimum Bounding Rectangle**, are stored. Instead, the `left_face` and `right_face` columns in the `edge_data` table are used to determine the geometry of the faces. Faces must not overlap or be contained within another face.

The data

In this chapter, a few data sources will be used as examples. All these sources are freely available on the Internet.

For the data model explanation and first processing examples, we will use the Natural Earth national borders data, specifically the **Admin 0 - countries** layer. It is available for download at `http://www.naturalearthdata.com/downloads/10m-cultural-vectors/`.

The data simplification topic will be discussed using the Czech hydrological dataset, called **DIBAVOD**. It is considered public information and is therefore free to download and reuse. The data can be downloaded from `http://www.dibavod.cz/index.php?id=27`. We'll need the following layer (download links are visible after expanding the *A* section): `A08 - hydrologické členění - povodí III.řádu` - the watershed dataset.

 There is also a more detailed version - the A07 dataset with fourth-order watersheds. It can be also used, but it will take more time to process.

Installation

The topology extension is not enabled automatically with basic PostGIS functionality. In order to use topology functions, they have to be activated using a PostgreSQL `CREATE EXTENSION` statement:

```
CREATE EXTENSION postgis_topology;
```

This will add topology functions to a database, and create metadata tables in it. Let's have a closer look at them.

We can see a newly created `topology` schema. It contains `topology` functions and two metadata tables:

- **Topology**: Storing information about separate topologies in a database
- **Layer**: Storing information about topological layers

A topology in PostGIS is a collection of topological elements (nodes, edges, and faces) with specified precision, coordinate system, and dimensionality (2D or 3D). Every topology is stored in a separate schema. A layer is a relationship between topology and a feature table. Each topology can have zero or more layers, and same topological elements can be used in more than one layer - which is useful for hierarchical data, such as administrative division or higher-order watersheds.

 As the topology features are encapsulated in separate schemas other than public, they aren't accessible to any unprivileged database user by default. It's necessary to grant USAGE and CREATE privileges on those schemas, and EXECUTE privileges on `topology` schema functions to any non-superuser who needs to use topology features.

Topology functions in PostGIS can be divided into two groups: those that are defined by the ISO standard, and those specific to PostGIS. Standard functions are prefixed with `ST_` while the non-standard functions aren't (this is a convention similar to one used by MS SQL Server). Sometimes, standard and non-standard functions have similar functionality, and the `ST_` variant exists purely for standard conformance. For detailed usage of topology functions, read on.

Creating an empty topology

After installing the topology extension, one more preparation step is needed before we can work with data: a new, empty topology has to be created. There are two functions for this: `topology.ST_InitTopoGeo` and `topology.CreateTopology`. The former is defined by the ISO standard, and accepts only one argument: the topology name. The latter is a non-standard version that allows the definition of additional parameters: the snapping tolerance, SRID, and dimensionality. Let's use the non-standard version:

```
SELECT topology.CreateTopology('my_topology', 4326, 0.00028, FALSE);
```

The first argument defines the topology name, the second is SRID (in this case, WGS84), the third specifies the precision (in this case, 1 arc-second, which is about 30 meters at the equator), and the fourth determines whether the topology should store Z-coordinates.

As a result, the function should return the integer ID of the newly created topology. Let's have a look at the database structure now:

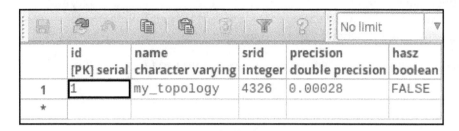

	id [PK] serial	name character varying	srid integer	precision double precision	hasz boolean
1	1	my_topology	4326	0.00028	FALSE
*					

Newly registered topology

In the topology schema, the newly created topology is registered in the topology.topology metadata table with parameters we have previously supplied as function arguments.

The new schema, called my_topology, has appeared. It contains the following tables: edge_data, face, node and relation.

It also contains an edge view, which contains a subset of the edge_data columns:

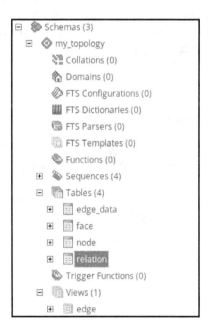

The database schema for topology

The relation table is a typical many-to-many relationship table, which connects topology elements with respective TopoGeometries in a feature table. The `element_type` column denotes the table where related elements reside: 1 is for node, 2 for edge, and 3 for face. Usually there's no need (and it's not recommended) to manipulate this table directly because the topology functions will take care of it.

These tables are now empty, as no data has been stored so far. We will populate them in the next section.

Importing Simple Feature data into topology

PostGIS can convert Simple Feature geometries into topology. Let's assume a countries table containing country boundaries from the Natural Earth dataset, with a geometry column named `geom`, and the WGS84 coordinate system.

Checking the validity of input geometries

The first step is to check the validity of input geometries. PostGIS will refuse to convert invalid geometries to topology elements, so it's very important to fix them or filter them out. We will use the `ST_IsValidReason` and `ST_IsValid` functions that we learned in `Chapter 1`, *Importing Spatial Data*:

```
SELECT ST_IsValidReason(geom) FROM countries WHERE ST_IsValid(geom) =
FALSE;
```

Luckily, in this case the query returned 0 rows, meaning that all geometries are valid, so we can proceed to the next step.

 It may happen that invalid geometries sneak into some revised version of Natural Earth, in which case the query will return a non-zero result. In that case, the invalid geometries must be repaired with the tools outlined in `Chapter 1`, *Importing Spatial Data*, or deleted before proceeding.

Creating a TopoGeometry column and a topology layer

`TopoGeometry` is a special data type. It doesn't store real geometry, merely an array of topological elements' IDs, but it can be visualized in GIS software, or used in spatial SQL queries, like an ordinary Simple Feature geometry, which will be composed on the fly. A new `TopoGeometry` column is created with a non-standard `topology.AddTopoGeometryColumn` function. It accepts the following arguments:

- `topology_name`: Denoting the name of topology which should be linked with the feature data
- `schema_name`: Name of schema where the feature table resides
- `table_name`: Name of the feature table
- `column_name`: Name for the newly created `TopoGeometry` column
- `geometry_type`: Geometry type for newly created `TopoGeometry`, one of: POINT, LINE, POLYGON, COLLECTION

For our example, the SQL statement will look like this:

```
SELECT
topology.AddTopoGeometryColumn('my_topology','public','countries','topogeom
','POLYGON');
```

After executing, a new row appears in the `topology.layer` table:

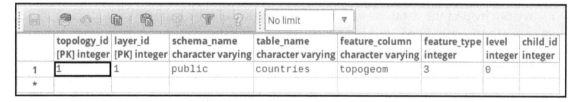

topology_id [PK] integer	layer_id [PK] integer	schema_name character varying	table_name character varying	feature_column character varying	feature_type integer	level integer	child_id integer
1	1	public	countries	topogeom	3	0	

A topology layer registered in a metadata table

Also, a new column, `topogeom`, is added to the `public.countries` table. But for now, all rows have NULL values there. To populate the newly created `TopoGeometry` column from an existing Geometry, another function has to be used.

Populating a TopoGeometry column from an existing geometry

The conversion function is called `topology.toTopoGeom`, and it accepts the following arguments:

- `geom`: Input geometry column
- `toponame`: Name of topology
- `layer_id`: ID of topology layer (it's returned by the `topology.AddTopoGeometry` function, and can be checked in the `topology.layer` table)
- `tolerance`: Snapping tolerance for conversion

The statement for our data will be as follows:

```
UPDATE countries SET topogeom =
topology.toTopoGeom(geom, 'my_topology',1,0.00028);
```

The conversion is a time-consuming process. For Natural Earth data, i7 CPU, 8 GB of RAM and a SSD, it took 6.5 minutes to complete. For bigger datasets, the process can take hours.

The conversion step can be used to fix some minor inconsistencies in non-topological data: unconnected lines or silver polygons. The errors will be corrected up to the tolerance parameter, that is, in our case, if lines are closer than 1 arc-second apart, they will become connected.

Now we can inspect the database again. The topological element tables (`nodes`, `edge_data`, and `faces`) were populated, and can be visualized in QGIS.

Note that, QGIS has a handy tool called **TopoViewer**. To use it, open DB Manager from the database main menu, select a `my_topology` schema, and pick TopoViewer from the Schema menu. This will add all topology elements to the map view, show the edge directionality, and label the elements' IDs:

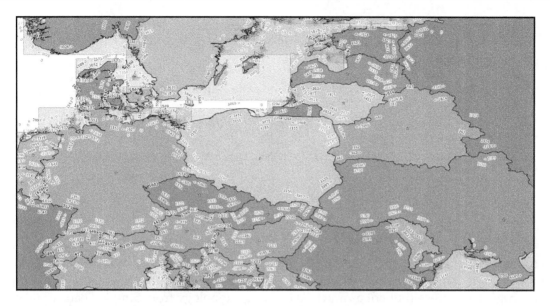

Topology visualized with QGIS TopoViewer

If everything is correct, it's time to DROP an existing Simple Feature geometry column: it will not be updated as the underlying topology changes. Instead, the new TopoGeometry column can be used (but not edited) in GIS software as though it was an ordinary geometry column, thanks to PostgreSQL's autocast feature:

```
ALTER TABLE countries DROP COLUMN geom;
```

Inspecting and validating a topology

After creating and populating a topology, the information is stored in multiple database tables. Luckily, PostGIS has a function for getting synthetic information about a given topology and its components: it's called topology.TopologySummary. It accepts one argument, the name of the topology in question:

```
SELECT topology.TopologySummary('my_topology');
                     topologysummary
---------------------------------------------------------------
Topology my_topology (id 2, SRID 4326, precision 0.00028)
4567 nodes, 4765 edges, 4257 faces, 255 topogeoms in 1 layers
Layer 1, type Polygonal (3), 255 topogeoms
 Deploy: public.countries.topogeom

(1 row)
```

In return, the function will print out the topology metadata (ID, SRID, and precision), the number of topology elements of each type, the number of layers, their ID, the geometry type, and the number of features in each layer.

Topology metadata can also be retrieved individually. There are three functions:

- `topology.GetTopologyID`: Given a topology name, returns its ID in a `topology.topology` table
- `topology.GetTopologySRID`: Given a topology name, returns its SRID
- `topology.GetTopologyName`: Given a topology ID, returns its name

Topology validation

A topology can be validated using the `topology.ValidateTopology` function. It takes one argument, the topology name. The computation is quite resource-hungry, so it can take some time to complete (but not as much as creating a topology from Simple Features takes).

The syntax for validating our topology is as follows:

```
SELECT topology.TopologySummary('my_topology');
```

After executing, if any errors are found, they are returned in a special composite type: `validatetopology_returntype`. This type consists of three parts:

- `error`, of type varchar, which contains a human-readable error message
- `id1`, of type integer, which contains the ID of the first problematic element
- `id2`, of type integer, which contains the ID of a second problematic element (if any)

Possible errors are as follows:

- Coincident nodes
- Edge crosses node
- Edge not simple
- Edge end node geometry mismatch
- Edge start node geometry mismatch
- Face overlaps face
- Face within face

Accessing the topology data

Elements used to compose TopoGeometries can be retrieved by manually querying the topology tables, but PostGIS provides specialized functions for that purpose. First, we will find topological elements of Poland using the `topology.GetTopoGeomElements` function:

```
SELECT topology.GetTopoGeomElements(topogeom) FROM countries WHERE
name='Poland';

 gettopogeomelements
---------------------
 {3155,3}
```

The function returns a set of rows, each one with one column of `topoelement` (which is just a two-element integer array) type. The first array element is a topological element ID, and the second is its type (1 - node, 2 - edge, 3 - face). Here we got one row, meaning the country is built from just one face.

For Norway, with lots of islands, the element list will be much longer:

```
SELECT topology.GetTopoGeomElements(topogeom) FROM countries WHERE name =
'Norway';
gettopogeomelements
---------------------
 {2713,3}
 {2714,3}
 {2724,3}
 ...
 {2831,3}
 {2832,3}
(120 rows)
```

There is also a similar function, `topology.GetTopoGeomElementArray`, which returns an array aggregate instead of set of rows.

To visualize the result, retrieving the elements' geometry is necessary. For area features, geometries can be retrieved as polygons or lines.

For polygons, the geometry is retrieved using the `topology.ST_GetFaceGeometry` function. It accepts two arguments, the topology name and face ID. In order to visualize all Norway's components as polygons, we will need the following query:

```
SELECT (topology.GetTopoGeomElements(topogeom))[1] AS face_id,
topology.ST_GetFaceGeometry('my_topology', (topology.GetTopoGeomElements(top
ogeom))[1]) FROM countries WHERE name='Norway';
```

The border of Norway visualized using topology.ST_GetFaceGeometry

Querying topological elements by a point

PostGIS has functions to identify topological elements at a given spatial location. There are three functions, each for one element type.

Locating nodes

First, we'll locate a node with the `topology.GetNodeByPoint` function. It's really a wrapper around `ST_Intersects` or `ST_DWithin` (when used with a tolerance parameter) functions.

It takes three arguments. The first is a topology name, the second is a point geometry (which must have the same SRID as topology), and the third is a tolerance (in the units of topology's SRID). The tolerance can be set to zero, in which case the point geometry must precisely match the node location or 0 will be returned.

For example, let's locate the node ID of a Czech-German-Polish tripoint near the town of Zittau:

```
SELECT topology.GetNodeByPoint('my_topology','SRID=4326;POINT(14.82337
50.87062)',0.02);
 getnodebypoint
----------------
           1598
```

When we set the tolerance too low, and leave the point geometry with five-digit precision, the function will return 0 instead:

```
SELECT topology.GetNodeByPoint('my_topology','SRID=4326;POINT(14.82337
50.87062)',0.002);
 getnodebypoint
----------------
              0
```

However, when the tolerance is set too high, there will be an exception, as this function can return only one node ID and no sorting by distance is done:

```
SELECT topology.GetNodeByPoint('my_topology','SRID=4326;POINT(14.82337
50.87062)',1);
ERROR:  Two or more nodes found
```

Locating edges

The next function, `GetEdgeByPoint`, is designed to find edges. Its use is identical, and also throws an exception when there is more than one edge within tolerance. When no edge is found within tolerance, 0 is returned:

```
SELECT topology.GetEdgeByPoint('my_topology','SRID=4326;POINT(14.99327
50.92510)',0.015);
 getedgebypoint
```

```
 ----------------
          1626

SELECT topology.GetEdgeByPoint('my_topology','SRID=4326;POINT(14.82061
50.87269)',0.02);
ERROR:  Two or more edges found
```

Another way to find edges in a topology is retrieving them by node. This time, more than one edge ID can be returned. Let's get back to our tripoint and use the `topology.GetNodeEdges` function:

```
SELECT topology.GetNodeEdges('my_topology', 1598);
 getnodeedges
--------------
 (1,-3470)
 (2,1626)
 (3,-1601)
```

This function returns a set of sequences: the edge appearance order (sequence) and the edge ID (edge).

Note that some IDs are returned as a negative value. This is because of edge direction:

```
SELECT edge_id, start_node,end_node FROM my_topology.edge_data WHERE
edge_id IN (3470,1626,1601);
 edge_id | start_node | end_node
---------+------------+----------
    1601 |       1574 |     1598
    1626 |       1598 |     3348
    3470 |       3351 |     1598
```

For edges returned with a negative ID, the node in question is an end node; for nodes returned with a positive ID, the node is a start node. When the edge is closed (for example, forming a polygon face) it will be returned twice, with both signs.

This is useful information, but if we want to retrieve the edges' geometry by joining, the result will have to be wrapped in the `abs()` function:

```
SELECT ed.edge_id, ed.geom FROM my_topology.edge_data ed JOIN
topology.GetNodeEdges('my_topology', 1598) ge ON abs(ge.edge) = ed.edge_id;
```

Locating faces

For faces, there is a `topology.GetFaceByPoint` function. Its usage is identical to the node and edge location functions: given a topology name, point geometry, and a tolerance, it returns a face ID:

```
SELECT topology.GetFaceByPoint('my_topology','SRID=4326;POINT(14.99327
50.92510)',0);
 getfacebypoint
----------------
           3155
```

When working with a contiguous boundary dataset, it's wise to set tolerance to zero, as when a point is close to the edge, more than one face can be found within tolerance and an error will be thrown.

Topology editing

Editing of a topological dataset is different than doing so with Simple Features, as the data is relational and spans across multiple tables. In this section, the editing workflow for PostGIS topology will be discussed.

Adding new elements

First, we will add new features to the topology using two different methods. One approach is to use the standard `ST_` functions. We will start with two isolated nodes. This is done with the `topology.ST_AddIsoNode` function. It takes three arguments: the topology name, the containing face ID (in our case, we will create nodes in empty space, so it will be `NULL`), and the point geometry:

```
SELECT topology.ST_AddIsoNode('my_topology',NULL,'SRID=4326;POINT(-1 1)');
 st_addisonode
----------------
           4568

SELECT topology.ST_AddIsoNode('my_topology',NULL,'SRID=4326;POINT(1 1)');
 st_addisonode
----------------
           4569
```

Note the returned nodes' IDs, as they will be necessary later when adding edges:

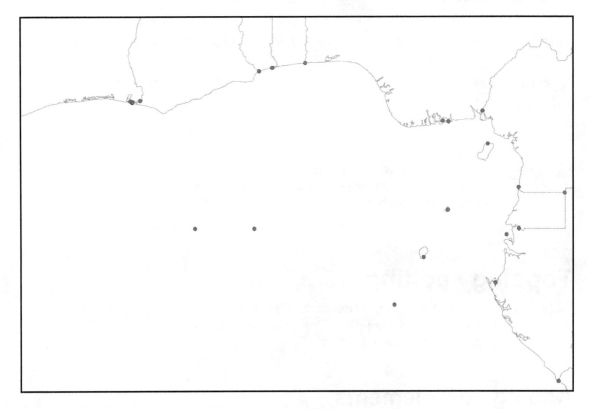

Two isolated nodes added to a topology

Next, it's time to add two edges, the first as isolated, the second creating a new face. The edge creation functions `ST_AddIsoEdge` and `ST_AddEdgeNewFaces` require four arguments: the topology name, the start node ID, the end node ID, and a `LINESTRING` geometry:

```
SELECT
topology.ST_AddIsoEdge('my_topology',4568,4569,'SRID=4326;LINESTRING(-1 1,
1 1)');
 st_addisoedge
---------------
          4766
```

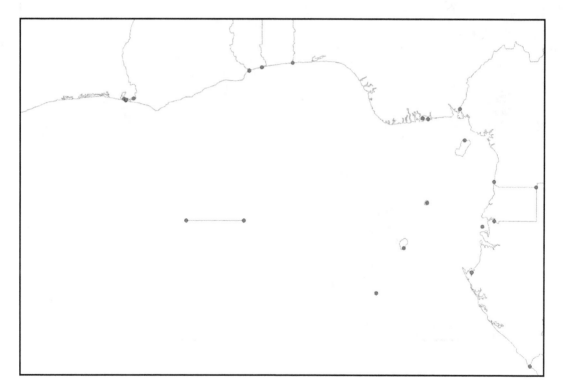

New isolated edge

We will close the polygon shape now. As the next edge will be bound to another edge, it cannot be called isolated, and therefore another function is necessary:

```
SELECT
topology.ST_AddEdgeNewFaces('my_topology',4569,4568,'SRID=4326;LINESTRING(1
1, 1 -1, -1 -1, -1 1)');
 st_addedgenewfaces
--------------------
        4767
```

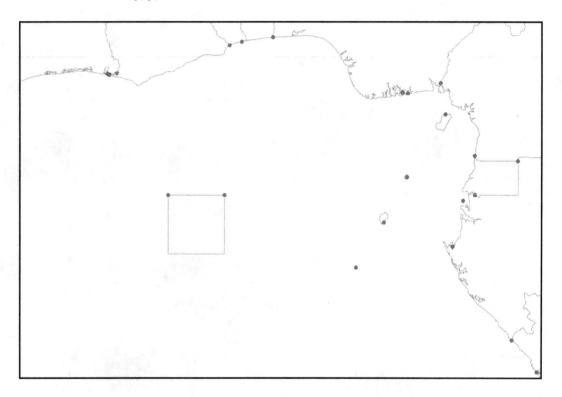

New closed edge

Now we can verify the newly added elements (and find out the face ID):

```
select edge_id, start_node, end_node, left_face, right_face
from my_topology.edge_data where edge_id in(4766,4767);
 edge_id | start_node | end_node | left_face | right_face
---------+------------+----------+-----------+------------
    4766 |       4568 |     4569 |         0 |       4258
    4767 |       4569 |     4568 |         0 |       4258
(2 rows)
```

The following is the face geometry:

```
SELECT ST_AsText(ST_GetFaceGeometry('my_topology',4258));
                st_astext
------------------------------------
 POLYGON((-1 1,1 1,1 -1,-1 -1,-1 1))
```

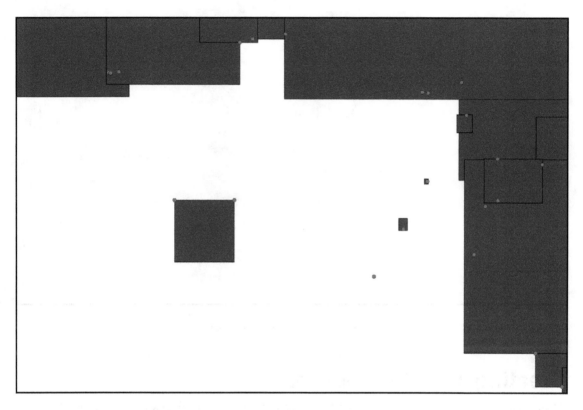

Newly created face

Another option is to use a PostGIS-specific function: `TopoGeo_AddPolygon`. When using it, we don't have to think about creating nodes and edges, as they will be created automatically. The function takes three arguments: the topology name, the polygon geometry (it has to be valid, and in the same SRID as topology), and a snapping tolerance when splitting existing edges:

```
SELECT topology.TopoGeo_AddPolygon('my_topology','SRID=4326;POLYGON((-2 1,
 -1.5 1, -1.5 -1, -2 -1, -2 1))',0.002);
 topogeo_addpolygon
--------------------
               4259
```

In this case, a new polygon was added in empty space. One closed edge has been created, as well as one node connecting its terminals. The newly created face ID was returned:

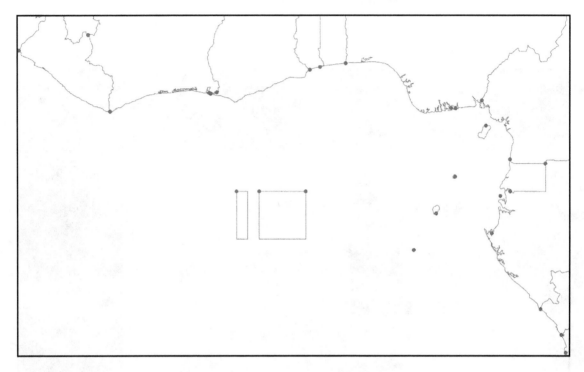

Newly created nodes and edges after the TopoGeo_AddPolygon function

Creating TopoGeometries

After creating some topological elements, it's time to link them with a feature table. A function, `topology.CreateTopoGeom`, is used to instantiate a `TopoGeometry`. It takes four arguments:

- Topology name
- Geometry type: `1 - Point`, `2 - Line`, `3 - Area`, `4 - Collection`
- `TopoGeometry` layer ID
- Array of TopoElements

Let's add the first created face as a fictional country, `Nulland`, to the countries table:

```
INSERT INTO countries(name,topogeom)
VALUES('Nulland',topology.CreateTopoGeom('my_topology',3,1,'{{4258,3}}'::to
pology.topoelementarray));
```

A `topoelementarray` is a special helper type. It's an array of two-dimensional arrays, each composed of a topology element ID and its type (1-node,2-edge,3-face). The element's types must match the `TopoGeometry` type if it's not a collection (type 4); for type 1, only nodes are allowed; for type 2, only edges; and for type 3, only faces.

The next face we create with `TopoGeo_AddPolygon` is another fictional country, `Neverland`, from the following:

```
INSERT INTO countries(name,topogeom)
VALUES('Neverland',topology.CreateTopoGeom('my_topology',3,1,'{{4259,3}}'::
topology.topoelementarray));
```

Splitting and merging features

In a topological data model, splitting and merging features is done by adding and removing nodes and edges. For example, a polygon will be split when an edge is added, and two lines are merged when a node between them is removed. In this section, we will learn how to merge and split features in PostGIS topology.

Splitting features

There are two ways to split a polygon - with SQL/MM standard function and a non-standard, PostGIS specific function.

Using the standard way, we'll have to add two nodes to existing edges and split them.

This is done with the `topology.ST_ModEdgeSplit` function. It accepts three arguments: the topology name, the edge ID, and the point geometry, and returns the integer ID of the newly created node. A new node is added to the topology, and the edge with a given ID is split into two parts. One retains the original ID, and the second is given a new ID.

> There is also a `ST_NewEdgeSplit` function, which deletes the original edge and creates two brand new edges instead. Use it if you are into immutable data structures.

We'll split `Nulland` into East and West `Nulland` by first splitting the edges with IDs of 4766 and 4767:

```
SELECT topology.ST_ModEdgeSplit('my_topology',4766,'SRID=4326;POINT(0 1)');
SELECT topology.ST_ModEdgeSplit('my_topology',4767,'SRID=4326;POINT(0
-1)');
```

The node and edge structure after adding two new nodes

Now, new nodes and edges are ready, but the face remains intact. So, we'll add a new edge, splitting the face - using the `topology.ST_AddEdgeModFace` function. It takes four arguments: the topology name, start node ID, end node ID, and new edge geometry. It splits a face, one part with the original ID remains in the database, and the second part is created. The return value is the ID of the newly created edge (the newly created face ID must be queried later):

```
SELECT
topology.ST_AddEdgeModFace('my_topology',4574,4575,'SRID=4326;LINESTRING(0
1, 0 -1)');
```

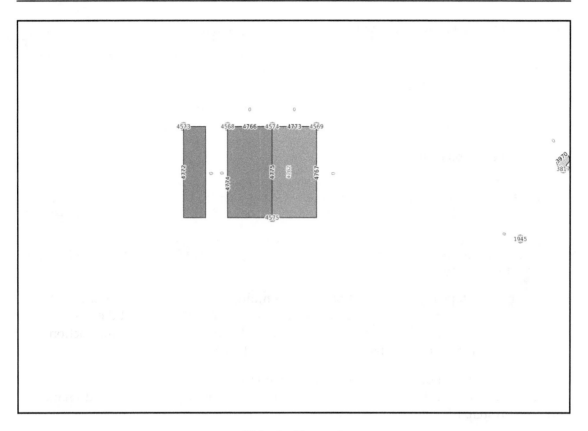

Split face after adding a new edge

 As with splitting an edge, there's ST_AddEdgeNewFaces, which creates two new faces instead of modifying existing ones.

Now the face is split, and `Nulland`'s `TopoGeometry` updated:

```
SELECT topology.GetTopoGeomElementArray(topogeom) FROM countries WHERE name
='Nulland';
 gettopogeomelementarray
-------------------------

 {{4258,3},{4262,3}}
```

Now it's time to update the feature table:

```
DELETE FROM countries WHERE name='Nulland';
INSERT INTO countries(name,topogeom) VALUES('West
Nulland',topology.CreateTopoGeom('my_topology',3,1,'{{4258,3}}'::topology.t
opoelementarray));
INSERT INTO countries(name,topogeom) VALUES('East
Nulland',topology.CreateTopoGeom('my_topology',3,1,'{{4252,3}}'::topology.t
opoelementarray));
```

As topology manipulation involves multiple queries, it's wise to use PostgreSQL's transactional features (`BEGIN TRANSACTION` at the beginning and `COMMIT` at the end). Should any query fail, the transaction ensures that the database is not left in a half-baked state.

Another way to split a polygon is to use a non-standard `topology.TopoGeo_AddLinestring` function. It creates the required nodes automatically, and can also snap the splitting line to a feature within a specified tolerance.

For example, in order to split `Neverland` into two pieces using a LineString geometry and `TopoGeo_AddLinestring` function, we need to execute a query as follows:

```
SELECT
topology.TopoGeo_AddLinestring('my_topology','SRID=4326;LINESTRING(-2.0001
0.001, -1.501 0.001)',0.002);
```

Note that the nodes and edges are added automatically, and the splitting line geometry doesn't have to be exactly snapped to an existing feature - it will be aligned within a specified tolerance:

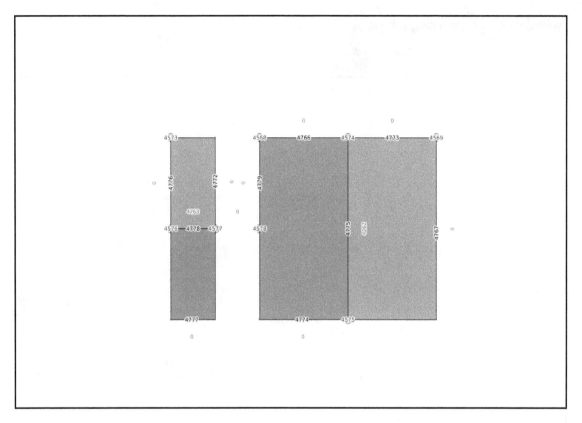

Split face after adding a LineString to a topology

Merging features

As mentioned before, merging faces is done by deleting an edge, and merging edges is done with node deletion. Let's assume West and East `Nulland` have reunited. There are two standard functions for edge removal: the mutating `topology.ST_RemEdgeModFace` and immutable `topology.ST_RemEdgeNewFaces`.

These functions don't work when any of the faces are members of any `TopoGeometry`, so we have to delete them first. For learning purposes, let's delete the whole `East Nulland` row and clear a `West Nulland`'s `TopoGeometry`:

```
DELETE FROM countries WHERE name='East Nulland';
UPDATE countries SET topogeom = topology.clearTopoGeom(topogeom) WHERE
name='West Nulland';
```

Now change the underlying topology:

```
SELECT topology.ST_RemEdgeModFace('my_topology',4775);
```

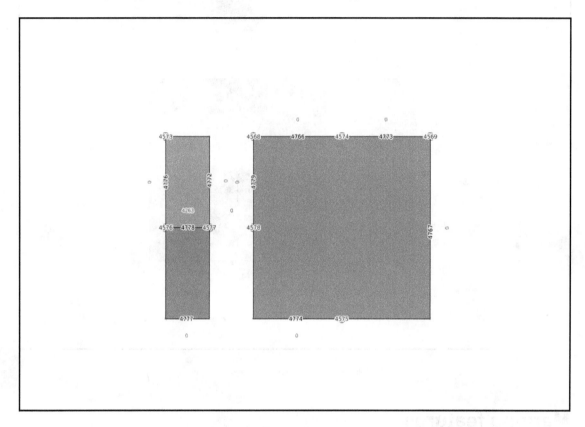

Merged faces after edge removal

The edge with ID 4775 is now removed, and the faces merged. The West
Nulland's TopoGeometry can be recreated:

```
UPDATE countries SET name = 'Nulland', topogeom =
topology.CreateTopoGeom('my_topology',3,1,'{{4258,3}}'::topology.topoelemen
tarray) WHERE name = 'West Nulland';
```

However, the two now-unnecessary nodes, `4574` and `4575`, remain. To get rid of them, we'll use the `topology.ST_ModEdgeHeal` function. It accepts three arguments: the topology name, the first edge ID, and the second edge ID. The first edge becomes modified with merged geometry, and the second one is deleted:

```
SELECT topology.ST_ModEdgeHeal('my_topology',4774,4767);
```

The ID of a deleted node (not edge!) is returned:

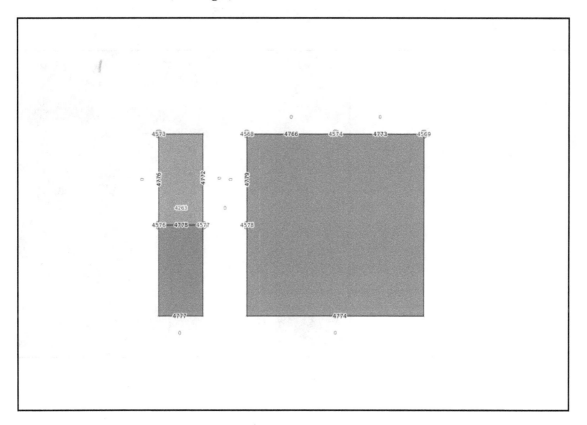

The edges merged with node removal

The second unnecessary node will be deleted when edges `4774` and `4773` are merged:

```
SELECT topology.ST_ModEdgeHeal('my_topology',4774,4773);
 st_modedgeheal
-----------------
           4569
```

Updating edge geometry

To modify the shape of a single edge, the `topology.ST_ChangeEdgeGeom` function is used. There are three arguments to supply: the topology name, the edge ID, and the new geometry. For example, we'll modify the border between East and West `Neverland`: the edge with an ID of `4778`:

```
SELECT
topology.ST_ChangeEdgeGeom('my_topology',4778,'SRID=4326;LINESTRING(-2
0.00099999999999989, -1.75 0.1, -1.5 0.001)');
```

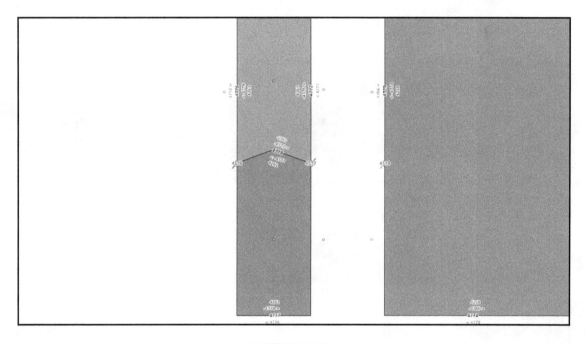

A modified edge geometry

Note that there's no tolerance, and the start and end points must precisely match. It's best to derive the first and last points for a new geometry from an existing geometry.

This is where PostGIS topology really shines: the geometry is changed once and only once, and both polygons will be still perfectly adjacent.

Topology-aware simplification

A very useful feature of a topology is the ability to simplify the feature geometry while maintaining its topological relationships. As we learned in Chapter 2, *Spatial Data Analysis*, the ST_SimplifyPreserveTopology function doesn't do that, despite its name. In this section, we will learn how to simplify features using topology functions.

Importing sample data

The data used as an example is extracted from the Czech hydrological dataset. Its location was mentioned at the beginning of the chapter, in the *The data* section. Now we'll import a DIBAVOD watershed layer into a database using ogr2ogr:

```
ogr2ogr -t_srs EPSG:32633 -f PostgreSQL "PG:dbname=mastering_postgis
host=localhost user=osm password=osm" -lco GEOMETRY_NAME=geom -lco
PRECISION=no -nln watershed_ord3  A08_Povodi_III.shp
```

Next, we will convert the Simple Feature geometry to a topology with 2-meter precision:

```
SELECT topology.CreateTopology('water_topology', 32633, 1, FALSE);
SELECT
topology.AddTopoGeometryColumn('water_topology','public','watershed_ord3','
topogeom','POLYGON');
UPDATE watershed_ord3 SET topogeom =
topology.toTopoGeom(geom,'water_topology',1,2);
```

We learned about geometry simplification in Chapter 2, *Spatial Data Analysis*. Let's try to use it to simplify watersheds with a 50-meter tolerance:

```
SELECT ogc_fid, ST_SimplifyPreserveTopology(geom,50) FROM watershed_ord3;
```

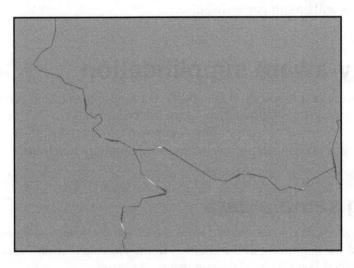

Simplification of a non-topological dataset

This is definitely not hydrologically correct: some areas belong to two watersheds and there are gaps.

With topology, it is possible to simplify a dataset without sacrificing its topological integrity. The trick is to use the topology.ST_Simplify function (not public.ST_Simplify) on a TopoGeometry column:

```
SELECT ogc_fid, topology.ST_Simplify(topogeom,50) from watershed_ord3;
```

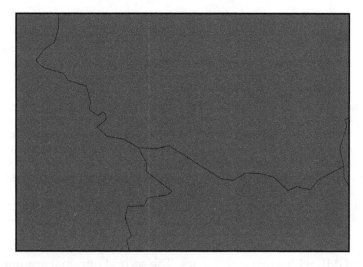

Simplification of a topological dataset

As result, geometries are simplified, but nodes are preserved and connectivity is not lost.

Topology output

There are two output functions in PostGIS Topology: `topology.ST_AsGML` and `topology.ST_AsTopoJSON`. **GML** is an OGC standard, while **TopoJSON** is a GeoJSON-like text-based exchange format. It was designed to reduce the size and redundancy of GeoJSON for datasets composed of neighboring polygons.

GML output

Topological GML can be retrieved from the database using the `topology.AsGML` function. The simplest use case is to supply a single argument, the `TopoGeometry` column:

```
SELECT topology.AsGML(topogeom) FROM countries WHERE name='Nulland';

<gml:TopoSurface><gml:directedFace><gml:Face
gml:id="F4258"><gml:directedEdge orientation="-"><gml:Edge
gml:id="E4766"><gml:directedNode orientation="-"><gml:Node
gml:id="N4568"/></gml:directedNode><gml:directedNode><gml:Node
gml:id="N4574"/></gml:directedNode><gml:curveProperty><gml:Curve
srsName="urn:ogc:def:crs:EPSG::4326"><gml:segments><gml:LineStringSegment><
gml:posList srsDimension="2">-1 1 0
1</gml:posList></gml:LineStringSegment></gml:segments></gml:Curve></gml:cur
```

```
veProperty></gml:Edge></gml:directedEdge><gml:directedEdge orientation="-
"><gml:Edge gml:id="E4774"><gml:directedNode orientation="-"><gml:Node
gml:id="N4574"/></gml:directedNode><gml:directedNode><gml:Node
gml:id="N4578"/></gml:directedNode><gml:curveProperty><gml:Curve
srsName="urn:ogc:def:crs:EPSG::4326"><gml:segments><gml:LineStringSegment><
gml:posList srsDimension="2">0 1 1 1 1 -1 0 -1 -1 -1 -1
0.001</gml:posList></gml:LineStringSegment></gml:segments></gml:Curve></gml
:curveProperty></gml:Edge></gml:directedEdge><gml:directedEdge
orientation="-"><gml:Edge gml:id="E4779"><gml:directedNode orientation="-
"><gml:Node gml:id="N4578"/></gml:directedNode><gml:directedNode><gml:Node
gml:id="N4568"/></gml:directedNode><gml:curveProperty><gml:Curve
srsName="urn:ogc:def:crs:EPSG::4326"><gml:segments><gml:LineStringSegment><
gml:posList srsDimension="2">-1 0.001 -1
1</gml:posList></gml:LineStringSegment></gml:segments></gml:Curve></gml:cur
veProperty></gml:Edge></gml:directedEdge></gml:Face></gml:directedFace></gm
l:TopoSurface>
```

This will output a GML with a `gml` namespace. The second optional argument can be given, and PostGIS will output XML elements in a custom namespace, such as `topogml`:

```
SELECT topology.AsGML(topogeom, 'topogml') FROM countries WHERE
name='Nulland';
```

```
<topogml:TopoSurface><topogml:directedFace><topogml:Face
topogml:id="F4258">...
```

Alternatively, we can create a GML document without namespaces, in which case we'll supply an empty string as the second argument:

```
SELECT topology.AsGML(topogeom, '') FROM countries WHERE name='Nulland';
```

```
<TopoSurface><directedFace><Face id="F4258"><directedEdge orientation="-
"><Edge id="E4766">...
```

TopoJSON output

`TopoJSON` is another topological format, designed with web mapping in mind. It is supported directly by the D3 and OpenLayers (starting with version 3) frameworks, and can be converted to GeoJSON with a TopoJSON library for use with other tools.

One difficulty with this format is that the edges (called arcs in TopoJSON parlance) are defined by index, while in PostGIS topology they are defined by IDs. To overcome this, an intermediate table called `edgemap` has to be created:

```
CREATE TABLE edgemap(arc_id serial, edge_id int unique);
```

> This table can be temporary, so PostgreSQL will automatically drop it at the end of a session.

Now the function is ready to be used. It accepts two arguments: the `TopoGeometry` column and the name of the intermediate table:

```
SELECT topology.astopojson(topogeom,'edgemap') FROM countries WHERE
name='Nulland';

                     astopojson
----------------------------------------------------
 { "type": "MultiPolygon", "arcs": [[[2,1,0]]]}
```

The intermediate table will have its rows populated with arc indexes and edge IDs:

```
SELECT * FROM edgemap;
 arc_id | edge_id
--------+---------
      1 |    4774
      2 |    4766
      3 |    4779
```

Now the TopoJSON document can be composed using the fragments generated by the `topology.AsTopoJSON` function and edge geometries. An example, using SQL functions only (no processing in the application layer) was shown in Chapter 7, *PostGIS - Creating Simple WebGIS Applications*.

Summary

PostGIS can store and analyze vector data not only in the Simple Feature model, but also in a topological model. This model is a good fit for data such as networks and boundaries, where connectivity and topological consistency are important. The toolset available is not as rich as Simple Feature's toolset, but all common processing tasks, such as creating, editing, merging, and splitting, can be accomplished with available functions. A very useful feature is the ability to simplify features while maintaining the topological integrity. Also, just by importing data into topology, some minor inconsistencies, such as unconnected lines, can be automatically fixed.

9
pgRouting

pgRouting is a PostGIS extension that brings routing tools to the table. pgRouting offers an extensive set of algorithms to choose from, can solve traveling salesman problems, calculate drive time zones, and even obey turn restrictions and avoid one-way streets.

With such a toolbox, one can create some pretty serious routing services that can be consumed, for example, by web applications.

In this chapter, we will focus on:

- Installing the pgRouting extension
- Importing routing data:
    - Importing shapefiles
    - Importing OSM data using `osm2pgrouting`
- Routing algorithms:
    - All pairs shortest path
    - Shortest path
    - Driving distance
    - Traveling sales person

Our final example will be a simple web application that calculates shortest routes in the city of Vienna.

Installing the pgRouting extension

Many distributions of PostGIS are equipped with pgRouting already, so there is a chance you have it installed without even knowing. Execute the following SQL to check whether you have pgRouting onboard:

```
select pgr_version();
```

If you do not get an error, but info on the pgRouting version, then you're good to go.

If you happen to not have the extension available, you need to obtain it first. To do so, navigate to `http://pgrouting.org/download.html` and follow the instructions for your OS. Once you have the binaries set up, enable the extension by executing the following:

```
CREATE EXTENSION pgrouting;
```

At this stage, we should be ready to continue our journey with pgRouting.

Importing routing data

In order to perform some routing analysis, we need the data first. You may obtain the data from different sources, we will use two of them - OSM data delivered in SHP format and OpenStreetMap data.

We will store the data in a new schema - pgr.

pgr is a prefix used by the pgRouting functions, so our schema fits perfectly in the naming convention.

Importing shapefiles

In this example, we will use an example of a routable shapefile downloaded from GeoFabrik.de - `https://www.geofabrik.de/data/shapefiles_routable_vienna.zip`. A routable shapefile, as GeoFabrik describes it, is a standard shapefile that contains OSM data preprocessed with routing in mind. This means it contains only road data with lines split at intersections, with some speed limits information, road line lengths, and so on.

We have already addressed importing shapefiles to PostGIS, so you can use a tool of your choice; in this case, I am using osm2pgsql:

```
shp2pgsql -s 4326 roads pgr.shp_roads | psql -h localhost -p 5434 -U
postgres -d mastering_postgis
```

Once our vector makes it to the database, we need to do some further processing before it is possible to issue pgRouting-specific queries against the dataset.

pgRouting requires the network edges (road lines) to be LineStrings. You may have noticed that shp2pgsql imported our lines as MultiLineStrings, but let's verify this:

```
select distinct st_GeometryType(geom) from pgr.shp_roads;
```

The result should be `ST MulitLineString` and we have to fix that.

 shp2pgsql does offer a -S param to create simple geometries instead of the default multi geometries. Unfortunately, our dataset of choice does contain multi geoms so we have to approach it differently.

In order to turn the MultiLineStrings into LineStrings (and split them when required) a `ST_Dump` function will come in handy. Basically, it extracts the geometry paths off a MultiGeometry:

```
select
    gid, osm_id, code, fclass, name, ref, oneway, maxspeed, layer, ete,
speed, length, bridge, tunnel,
    (ST_Dump(geom)).geom as geom
into pgr.shp_roads_fixed
from
    pgr.shp_roads;
```

Let's verify our fixed data - this time, the expected output is `ST LineString`:

```
select distinct st_GeometryType(geom) from pgr.shp_roads_fixed;
```

pgRouting works with its own specific flavor of topology, so we need to create pgRouting topology for our data. We will use a `pgr_CreateTopology` function to prepare the data. It requires two extra columns in the roads dataset, both integers - one for the start point identifier and the other one for the endpoint identifier; their default names are source and target, respectively. Let's modify our roads data model to cater for the requirements:

```
ALTER TABLE pgr.shp_roads_fixed ADD COLUMN source integer;
ALTER TABLE pgr.shp_roads_fixed ADD COLUMN target integer;
```

At this stage, we should be able to create the pgRouting topology. So, let's do just that:

```
select pgr_createTopology(
    'pgr.shp_roads_fixed', --edge_table; network table name
    0.000001, --tolerance; snapping tolerance with disconnected edges
    'geom', --name of the geometry column in the network table
    'gid', --name of the identifier column in the network table
    'source', --name of the source identifier column in the network table
    'target' --name of the target identifier column in the network table
    --rows_where: condition to select a subset of records; defaults to true
to process all the rows where the source / target are nulls; otherwise the
subset of rows is processed
    --clean: boolean - clean any previous topology; defaults to false
);
```

In the preceding code, I included the explanation of each parameter. Basically, this code creates pgRouting topology for our roads dataset. The roads table gets modified - source and target columns are filled with new data; indexes are created for ID, geom, source, and target columns.

If the topology has been created, you should see an OK result; FAIL otherwise. If the operation failed and you happen to use pgAdmin to run queries, review the Messages tab for details on what the reason for the failure.

You should notice that a new table has been created: shp_roads_fixed_vertices_pgr.

We can now verify the graph by using the pgr_analyzeGraph function:

```
select pgr_analyzeGraph('pgr.shp_roads_fixed', 0.00001, 'geom', 'gid');
```

It will also show either OK or FAIL, but when you look at the notifications (if you run the command via psql, you will see the output in the CMD; in pgAdmin you will need to switch to the **Messages** tab) you should see a similar output:

```
NOTICE:  Performing checks, please wait ...
NOTICE:  Analyzing for dead ends. Please wait...
NOTICE:  Analyzing for gaps. Please wait...
NOTICE:  Analyzing for isolated edges. Please wait...
NOTICE:  Analyzing for ring geometries. Please wait...
NOTICE:  Analyzing for intersections. Please wait...
NOTICE:              ANALYSIS RESULTS FOR SELECTED EDGES:
NOTICE:                   Isolated segments: 59
NOTICE:                          Dead ends: 7534
NOTICE:  Potential gaps found near dead ends: 17
NOTICE:              Intersections detected: 1064
NOTICE:                    Ring geometries: 220
 pgr_analyzegraph
```

```
------------------
 OK
(1 row)
```

Dead ends and potential gap problems are also identified in the vertices table in the `cnt` and `chk` columns:

```
select
    (select count(*) FROM pgr.shp_roads_fixed_vertices_pgr WHERE cnt = 1) as
deadends,
    (select count(*) FROM pgr.shp_roads_fixed_vertices_pgr WHERE chk = 1) as
gaps
```

Importing OSM data using osm2pgrouting

If you happen to prefer OpenStreetMap data for your routing solutions, you can use the `osm2pgrouting` utility to import network data directly from the OSM format.

As mentioned earlier, pgRouting may have been bundled with your PostGIS installation. In such cases, `osm2pgrouting` should already be installed - see if the utility is available in the PgSQL's bin, or simply type `osm2pgrouting` in the console.

If you see help for the tool, you're good to go; otherwise, you will have to install it.

`osm2pgrouting` gets much more love from Unix users (obviously), so if you happen to be one of a kind, in order to get the utility, follow the instructions from here: `https://github.com/pgRouting/osm2pgrouting/tree/osm2pgrouting-2.1.0`.

Windows versions of `osm2pgrouting` are available under the *Unreleased PostGIS Versions* section of the PostGIS download website (`http://postgis.net/windows_downloads/`) - do make sure you download the version appropriate for your setup - make sure the database version and PostGIS version match. Once downloaded, follow the instructions in the readme.

For this example, we will use data for Vienna from MapZen's metro extracts: `https://s3.amazonaws.com/metro-extracts.mapzen.com/vienna_austria.osm.bz2`. This is the `Raw OpenStreetMap dataset` in XML format.

Once downloaded and extracted, we can import it:

```
osm2pgrouting --conf <mapconfig.xml path> --file vienna_austria.osm --
schema pgr --clean 1 --host localhost --db_port 5434 --user postgres --
passwd postgres --dbname mastering_postgis
```

 If you are on a Linux box, you can omit the -conf parameter; the tool will use the default mapconfig.xml. On a Windows box, you need to provide the path to the PgSQL's bin, or execute the tool from this directory.

When you review the pgr schema, you will notice that there are a few more tables now - a couple of OSM dictionaries and ways and ways_vertices_pgr tables.

The data model of the ways table differs from the model imported from SHP. It is worth knowing the meaning of some of the data:

- length: Length of an edge in degrees
- length_m: Length of an edge in metres
- cost/reverse_cost: Length of an edge in degrees; negative value is for the wrong way
- cost_s/reverse_cost_s: Time in seconds calculated for the max speed for an edge

pgRouting algorithms

pgRouting is equipped with quite a few algorithms specialized in different aspects of routing. Although all are routing algorithms, for the sake of convenience, let's split them into the following artificial functional groups:

- All pairs shortest path
- Shortest Path
- Driving distance
- Traveling Sales Person

All pairs shortest path

All pairs shortest path algorithms are good for calculating the total costs of the shortest path for each node in the graph. There are two of those available in pgRouting:

- **All pairs shortest path, Johnson's algorithm**: Good for calculating costs over sparse graphs
- **All pairs shortest path, Floyd-Warshall algorithm**: Good for calculating costs over dense graphs

Let's give it a shot:

```
select * from pgr_floydWarshall('select gid as id, the_geom, source::int4,
target::int4, cost::float from pgr.ways where ST_Intersects(the_geom,
ST_MakeEnvelope(16.3618,48.2035,16.3763,48.2112,4326))', false);
```

This query picks roughly a thousand of the edges of Vienna city center from our OSM ways table and outputs something like this (the documentation mentions that the recommended usage is running the method on up to 3500 rows, as otherwise the performance will drop drastically):

start_vid	end_vid	agg_cost
58529	68104	0.000602242417967414
58529	47346	0.00909508063542836
58529	48852	0.00986963261560741
58529	39868	0.00571458698714135
58529	40505	0.00620416841040188

. . .

All pairs algorithms do not return a path; what you get is a list of total costs of the shortest paths between each node of the graph. The first row reports cost 0, as the start (id1) and end (id2) nodes are the same. Next goes the list of costs for the shortest routes to the other nodes, starting from the node in the id1 column.

Our query returns over a million combinations, so it may take a while to compute.

In version 2.2, the algorithms mentioned in this section were redesigned to improve performance; method names and their signatures, as well as the output, changed a bit. If you use pgr older than 2.2, you should look for the `pgr_apspwarshall` function instead.

Shortest path

Shortest path algorithms are tasked with calculating the shortest path between specified points. pgRouting offers the following to choose from:

- Shortest path Dijkstra
- Bi-directional Dijkstra shortest path
- Shortest path A*
- Bi-directional A* shortest path
- K-shortest path, multiple alternative paths
- K-Dijkstra, One-to-Many Shortest Path
- Turn restrictions shortest path (TRSP)

Shortest path does not necessarily mean shortest in terms of length. If a network has some other characteristics assigned, for example, time needed to travel an edge, or the cost of travel, the output can be considered the quickest or cheapest respectively. The meaning of the 'cost' provided to the shortest path algorithms can therefore be adjusted to suit even sophisticated needs and calculate routes with the lowest denivelation, for example.

The output of the algorithms is not geometry as one could expect, but rather a dataset that describes the sequence of nodes and edges, along with the cost of traveling through an edge and the cumulative cost of a route. We'll see some examples a bit later.

You may have noticed that both Dijskatra and A* algorithms have bi-directional variants. In such cases, the path is being calculated simultaneously from both ends (source and target) and stopped in between whenever the paths meet. This is meant to improve performance.

Shortest path Dijkstra

The algorithm was invented in 1956 by Edsger W. Dijkstra, a computer scientist tasked to demonstrate the capabilities of the ARMAC computer while working at the Mathematical Center in Amsterdam.

Shortest path Dijkstra was the first algorithm implemented in pgRouting; pgRouting was called **pgDijkstra** in its early days.

The algorithm can be used by calling a `pgr_dijkstra` function. It has a few different signatures that let one calculate paths in the following **modes**:

- **One-to-one**: One source node to one target node
- **One-to-many**: One source node to many target nodes
- **Many-to-one**: Many source nodes to one target node
- **Many-to-many**: Many source nodes to many targets

All the methods can be used for either directed or non-directed networks.

The simplest usage of the algorithm looks like this:

```
select * from pgr_dijkstra('select gid as id, source, target, length_m as
cost from pgr.ways', 79240, 9064);
```

This uses the minimal signature of **pgr_dijkstra** and calculates the route from point A (in our case, Schottenfeldgasse) to point B (in our case, Beatrixgasse) and assumes a directed graph. The query results in 228 rows and the result looks as follows:

```
 seq|  path_seq|   node|  edge|               cost|          agg_cost
-----+----------+-------+-------+------------------+------------------
  1|         1|  79240|  69362|  169.494690094003|                 0
  2|         2|  52082|  44175|  107.754808368861|  169.494690094003
...
```

In order to obtain the actual geometry, we should modify the query a bit:

```
select
   ways.the_geom
from (
   select * from pgr_dijkstra('select gid as id, source, target, length_m
as cost from pgr.ways', 79240, 9064)    ) as route
   left outer join pgr.ways ways on ways.gid = route.edge;
```

When visualized, our result looks as follows:

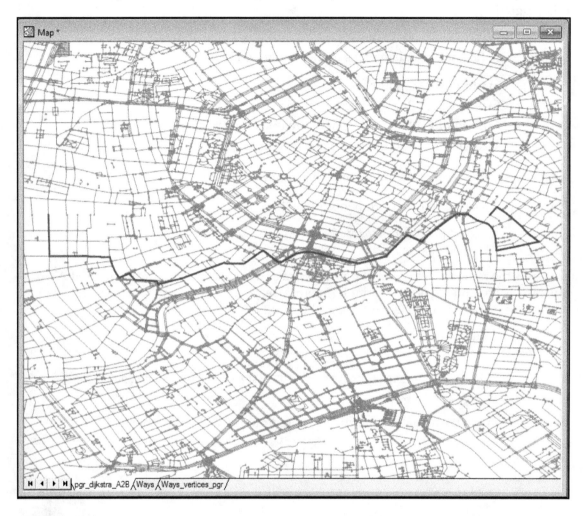

The output of the algorithm in the many mode is a bit different; the returned result set adds two extra columns (depending on the scenario) - start_vid and end_vid to specify the starting and/or the ending node, respectively. Let's use our previously used locations to search for the shortest path to Vienna's Central Train station:

```
select * from pgr_dijkstra('select gid as id, source, target, length_m as
cost from pgr.ways', ARRAY[79240, 9064], 120829);
```

The results are very similar, but you will notice the presence of a `start_vid` column, described earlier:

```
    seq|path_seq|start_vid|   node|  edge|       cost|        agg_cost
 -----+--------+---------+-------+-------+-----------+----------------
     1|       1|    9064|    9064| 209769|93.17915259|0
     2|       2|    9064|   15615| 133330|82.53885472|93.179152510029
 . . .
```

This time, our routes are as follows (the route between both points is also shown on the screenshot):

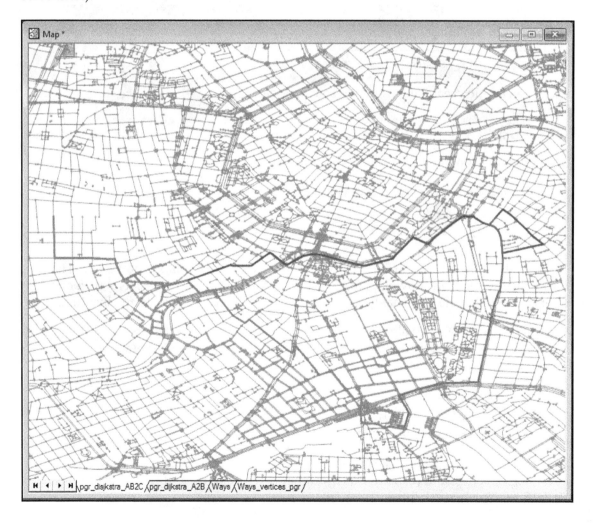

An excellent illustration of how the Dijkstra algorithm works can be found on Wikipedia at:
`https://en.wikipedia.org/wiki/File:Dijkstras_progress_animation.gif`.

If you happen to only need the cost of the shortest path and you do not need the information about the vertices and edges per se, you can use the `pgr_dijsktraCost` function. It has the same signatures as the regular `pgr_dijkstra`, but returns simplified results:

```
select * from pgr_dijkstraCost('select gid as id, source, target, length_m
as cost from pgr.ways', 79240, 9064);
```

start_vid	end_vid	agg_cost
79240	9064	5652.40083087719

And for a variation searching from two source nodes:

```
select * from pgr_dijkstraCost('select gid as id, source, target, length_m
as cost from pgr.ways', ARRAY[79240, 9064], 120829);
```

start_vid	end_vid	agg_cost
9064	120829	2860.06209537015
79240	120829	4532.12388804859

A-Star (A*)

The algorithm was invented in 1968 by Nils Nilsson, Bertram Raphael, and Peter E. Hart during their work on improving path planning of Shakey the robot - a general purpose robot developed at the Artificial Intelligence Center of the Stanford Research Institute.

A* is similar to the Dijkstra algorithm in that it searches among all the possible paths. The improvement is that it uses heuristics to first consider the paths that seem to lead to the target quicker.

The basic usage is as follows:

```
select * from pgr_aStar('select gid as id, source, target, cost, x1, y1,
x2, y2 from pgr.ways', 79240, 9064);
```

This time, we need to provide a bit more data to the algorithm; not only the edge ID, source vertex, target vertex, and cost, but also coordinates of the source and target vertices needed for the *heuristic magic* to happen.

When heuristics is not used, the A* algorithm behaves the same way as Dijkstra.

The output is exactly what we saw when using one-to-one Dijkstra:

```
seq|  path_seq|   node|   edge|              cost|          agg_cost
----+----------+-------+-------+------------------+------------------
  1|         1|  79240|  69362| 169.494690094003|                 0
  2|         2|  52082|  44175| 107.754808368861| 169.494690094003
. . .
```

And the found route is also exactly the same:

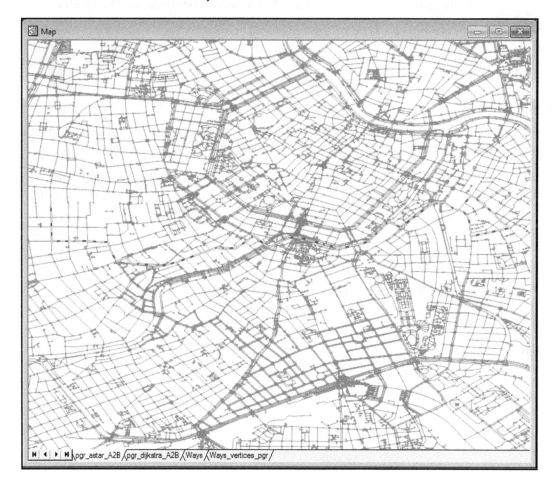

We searched for the shortest path between the same vertices, so we can draw some assumptions by now. Basically, A* is quicker than Dijsktra, but the difference depends on the data and the dataset size. On my box, calculating Dijsktra for our test points takes roughly 470 ms, while A* takes 450 ms.

This is by no means a trustworthy benchmark of course, but it gives a general impression of what to expect.

It is also worth remembering that, since A* star is a best-fit algorithm, it tries to predict what vertices to evaluate in order to find the shortest path; it may not return the most optimal result. At the same time, Dijkstra evaluates more data and guarantees the results to be optimal.

 Some good illustrations of how A* works can be found on Wikipedia at `ht tps://en.wikipedia.org/wiki/A*_search_algorithm`.

K-Dijkstra

This algorithm is designed to calculate paths from one source to multiple targets and it is represented by two functions: `pgr_kdijkstraPath` and `pgr_kdijkstraCost`.

As of pgRouting 2.2, K-Dijkstra is deprecated and its functionality is provided by the `pgr_dijkstra` set of functions. Since we have already seen this in action (although in a reverse mode - many sources to one target), there is no sense in repeating this again.

K-Shortest path

This algorithm is designed to return multiple alternative paths, so it not only finds the shortest path, but also K-1 alternative paths. pgRouting implementation is based on Jin J. Yen's works.

The basic usage is as follows:

```
select * from pgr_ksp('select gid as id, source, target, cost, x1, y1, x2, y2 from pgr.ways', 79240, 9064, 2);
```

The output is very similar to what we have seen so far; the difference is the presence of an additional column that indicates the path identifier:

```
seq|path_id|path_seq| node| edge|                cost|           agg_cost
---+--------+--------+-----+-----+--------------------+------------------
  1|      1|       1|79240|69362|  0.00152459527744|                  0
  2|      1|       2|52082|44175|0.000969276126804|0.001524595277445
. . .
```

When we visualize the preceding output, it looks as follows:

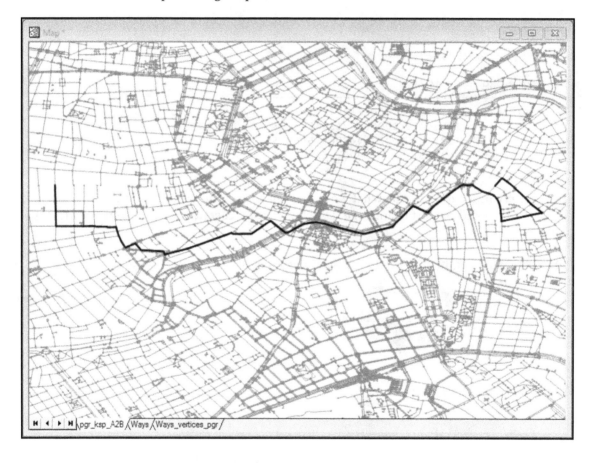

Turn restrictions shortest path (TRSP)

Turn restrictions shortest path is an algorithm that can make use of turn restrictions, so it possible to model real-world scenarios. In terms of performance, it should be close to A*.

For this example, we will focus on a smaller area to demonstrate how to define and use turn restrictions that should be fed to the `pgr_trsp` function.

Let's first calculate a route from vertex 26306 to vertex 98111:

```
select * from pgr_trsp('select gid::int4 as id, source::int4, target::int4,
length_m::float8 as cost from pgr.ways', 26306, 98111, false, false);
```

Our output is as follows:

```
Seq|   id1|    id2|              cost
----+------+-------+--------------------
   0| 26306| 41122| 0.00282144411605047
   1| 51491| 132340|0.000500071364908483
   2| 73162| 132341| 0.00714439682618376
   3| 26893| 157023| 0.00634007469987476
   4| 98111|    -1|                    0
```

This is pretty much the same as what we have seen so far. The meanings of the columns are: sequence, vertex ID to start from, edge ID to follow, and cost of traveling from node (id1) using edge (id2).

Now let's create the turn restriction table, as described in the documentation:

```
CREATE TABLE pgr.restrictions (
    id serial,
    to_cost double precision,
    target_id integer,
    via_path text
);
```

And define some restrictions:

```
INSERT INTO pgr.restrictions (to_cost,target_id,via_path) VALUES (10000,
157023, '132341');
```

The preceding code means that if you are traveling along edge `132341`, then the cost of traveling via edge `157023` is going to be `10000`.

Let's see if our restriction works:

```
select * from pgr_trsp('select gid::int4 as id, source::int4, target::int4,
cost::float8 as cost from pgr.ways', 26306, 98111, false, false, 'select
to_cost, target_id, via_path from pgr.restrictions');
```

And it looks like it does; the sequence of edges to travel has changed:

```
Seq|   id1|    id2|                   cost
----+------+-------+--------------------
  0| 26306|  41122| 0.00282144411605047
  1| 51491| 132340|0.000500071364908484
  2| 73162|  91344| 0.00321165171991199
  3| 39894| 136118| 0.00652205259101692
  4|  2717|  66313| 0.00394702020516963
  5| 26893| 157023| 0.00634007469987476
  6| 98111|    -1|                    0
```

It is possible to define more than one edge as the access path:

```
INSERT INTO pgr.restrictions VALUES (2, 10000, 157023, '66313,136118');
```

This time, what the restriction says is: if you are traveling via edges `136118` - `66313`, then the cost of continuing onto `157023` is `10000`.

 Notice the route chain that defines the access path to the restricted edge comes in reverse order.

Once again, the order of the edges to travel has changed. For a change, instead of showing the tabular data again, let's see how our routes look on a map:

- Orange route is the route without the turn restrictions; it does indeed look like the shortest path
- Yellow route was calculated with the first restriction: if coming from `132341`, then the penalty of traveling onto `157023` was `10000`

- Green route was calculated with two restrictions - the first one and another one says that traveling onto `157023` from `136118 - 66313` would incur a penalty of `10000`

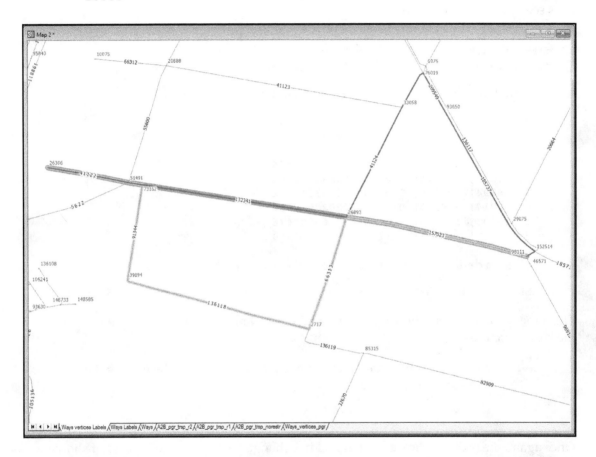

Driving distance

Driving distance functions use the Dijkstra algorithm to find all the edges that have the cumulative costs less or equal to the specified value. Our ways dataset has some properties that can be used to perform sensible driving distance analysis.

Let's first find the edges that are no farther than 1000 metres from Vienna's central train station:

```
select * from pgr_drivingDistance('select gid as id, source, target,
length_m as cost from pgr.ways', 120829, 1000);
```

Next, let's see how far we can get from the station in five minutes:

```
select * from pgr_drivingDistance('select gid as id, source, target, cost_s
as cost from pgr.ways', 120829, 300);
```

In both cases, the output is a dataset with the following columns: `seq`, `node`, `edge`, `cost`, and `agg_cost`. Query results visualized on a map look as follows (orange is the result of the first query and blue shows the results of the second one):

Pgr_drivingDistance can also accept multiple starting nodes, so it is possible to calculate more complex scenarios in one go. In such cases, start vertices are provided as an array, and the output gets one more column - from_v - that describes what the starting point for the data was.

Now, as we have calculated the driving distance along our network, let's create driving distance zones (also called drive time zones, catchment areas) for our five minute drive, so we can assess what area actually belongs to a certain driving distance. In order to do so, we will use a pgr_alphaShape function:

```
select pgr_alphaShape(
    'select
          v.id::int4, v.lon::float8 as x, v.lat::float8 as y
    from(
          select * from pgr_drivingDistance(''select gid as id, source,
target, cost_s as cost from pgr.ways'', 120829, 300)
       ) as dd
    left outer join pgr.ways_vertices_pgr v on dd.node = v.id'
);
```

The returned data is not a polygon as one could expect, but a set of pgr_alphashaperesult; so, basically, points that make up a shape:

```
       pgr_alphashape
-----------------------
(16.4133473,48.1908108)
(16.4131929,48.1906814)
...
(,) <-ring separator
...
```

If the function outputs more than one outer/inner ring, then they are separated by a row with null values for both x and y.

Next, let's make a polygon out of the calculated points:

```
select pgr_pointsAsPolygon(
        'select
              v.id::int4, v.lon::float8 as x, v.lat::float8 as y
        from(
              select * from pgr_drivingDistance(''''select gid as id,
source, target, cost_s as cost from pgr.ways'''', 120829, 300)
           ) as dd
        left outer join pgr.ways_vertices_pgr v on dd.node = v.id'
    );
```

Because `pgr_pointsAsPolygon` accepts text as the vertices select query and passes it internally to `pgr_alphaShape`, it may sometimes be difficult to work out how to escape apostrophes properly. In such cases, it is worth having a look at the PostgreSQL's double dollar sign delimiter: `$$`. In our example, it would be enough to replace the inner apostrophes with `$$`. If you happen to have more levels of string nesting, you add an identifier between the dollar symbols: `$lvl1$ some string $lvl2$ some nested quoted string $lvl2$ some text $lvl1$`

Finally, our result is a MultiPolygon:

Traveling sales person

The last pgRouting algorithm that we are about to have a look at is the traveling sales person algorithm. Its purpose is to answer the following question: Given a list of cities and the distances between each pair of cities, what is the shortest possible route that visits each city exactly once and returns to the origin city?

pgRouting offers two variants of a TSP function:

- `pgr_eucledianTSP`: Takes a set of points and calculates a point order from the internally calculated cost matrix based on the Euclidean distances calculated from point coordinates expressed as Lon/Lat
- `pgr_TSP`: Takes an already calculated distance matrix

Let's calculate the points order for a couple of locations from our Vienna dataset:

```
select * from pgr_eucledianTSP(
    'select id, lon::float8 as x, lat::float8 as y from
pgr.ways_vertices_pgr where id in
(4540,57407,116126,58791,44800,85852,52421,148735)'
);
```

The output is as follows:

```
 seq|   node|                cost|            agg_cost
-----+-------+--------------------+--------------------
   1|   4540| 0.0272292439764646|                   0
   2|148735| 0.0149659383701813|0.0272292439764646
   3| 57407|0.00798812567815924|0.0421951823466459
   4|116126| 0.0406742028363178|0.0501833080248051
   5| 85852| 0.0422062138944045| 0.090857510861123
   6| 58791| 0.0105530643293782| 0.133063724755527
   7| 44800| 0.0181016168186681| 0.143616789084906
   8| 52421| 0.0106700309212339| 0.161718405903574
   9|   4540|                  0| 0.172388436824808
```

The preceding output on a map looks as follows:

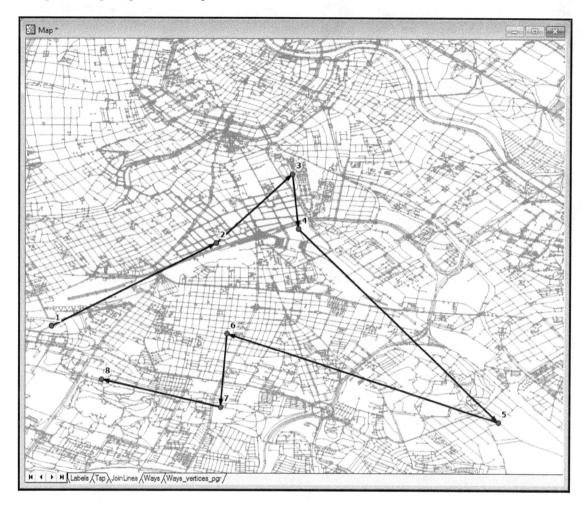

pgRouting 2.2 documentation mentions that using Dijkstra to pre-calculate aggregate costs of the shortest paths between a set of points prior to feeding the data to TSP is not worthy, as solving TSP on a Euclidean plane first and then finding shortest paths in an ordered set of points should give results that are fairly optimal already.

pgRouting 2.3, on the other hand, brings a `pgr_dijkstraCostMatrix` function, so one should expect that providing data with the distance matrix calculated with real distances is sensible enough to use it. Let's do just that:

```
select * from pgr_TSP(
    $$
    SELECT * FROM pgr_dijkstraCostMatrix(
        'SELECT gid as id, source, target, cost, reverse_cost FROM
pgr.ways',
        (select array_agg(id) from pgr.ways_vertices_pgr where id in
(4540,57407,116126,58791,44800,85852,52421,148735)), directed := false)
    $$,
    start_id := 4540, --if not specified, first in arr would be used
    randomize := false
);
```

Calculating distance matrix with directed networks will likely result in the output matrix being non-symmetric - the cost of traveling from A to B is not the same as the cost of traveling from B to A (cost/reverse cost). This will make the TSP function fail with the following message:
ERROR: A Non symmetric Matrix was given as input

The output is actually the same, although in reversed order:

seq	node	cost	agg_cost
1	4540	0.0156343305442904	0
2	52421	0.0255674722245925	0.0156343305442904
3	44800	0.0133600823412494	0.0412018027688828
4	58791	0.0458230476499673	0.0545618851101322
5	85852	0.0475591325608726	0.1003849327601
6	116126	0.0186155727831978	0.147944065320972
7	57407	0.0272063328584911	0.16655963810417
8	148735	0.0297395689283764	0.193765970962661
9	4540	0	0.223505539891037

Starting with pgRouting 2.3, there are quite a few extra optional parameters that can tweak the way TSP is solved. It is worth doing some further reading on the algorithm implementation though, to fully understand the impact they have on the calculation process.

Handling one-way edges

All the routing algorithms that we have seen have an option to specify whether a graph is directed or non-directed. An edge of non-directed graphs can be travelled from source to target and from target to source with the same cost. When the cost of traveling in both directions is different, or an edge is not traversable in one of the directions, then a network is directed.

In pgRouting, cost and reverse cost is used for defining the rules of traveling in both directions (source to target and target to source). So, whenever telling the algorithms to assume a directed network is used, the reverse cost should be provided.

Basically, the rules are as follows:

- Cost applies to edge traversal from source node to target node or, when a network is not directed, also from the target node to source node
- Reverse cost applies to edge traversal from target to source and will only be used when assuming a directed network
- To discourage the algorithm to use an edge in a given direction, cost should be set to a higher value
- To remove an edge in a given direction from a graph completely, a cost should be set to a negative value

For example:

```
id| source| target| cost| reverse_cost
----+-------+-------+-----+--------------
  1|      1|      2|  0.1|           0.1
  2|      1|      3|  0.1|         10000
  3|      2|      3|   -1|           0.1
  4|      2|      4|  0.1|            -1
```

- Edge 1 is traversable in both directions
- Edge 2 is also traversable in both directions, but the cost is high enough. So, if an edge has been used, it would mean there was no other way to get to the target; normally, algorithms would avoid traversing such edges
- Edge 3 is only traversable from target to source node (3 -> 2); edge connecting nodes 2 -> 3 is not considered a part of the graph
- Edge 4 is traversable only from source to target, but not the other way round; edge connecting nodes 4 -> 2 is not considered a part of the graph

Consuming pgRouting functionality in a web app

A final example of our pgRouting journey is a web application that consumes some of the functionality we have seen so far.

In order to preview the example, navigate to the example's folder - `apps/pgrouting`, run `sencha app watch`, and navigate to `http://localhost:1841/apps/pgrouting/`. You should see a similar output (you will have to calculate a route and drive time zone first though):

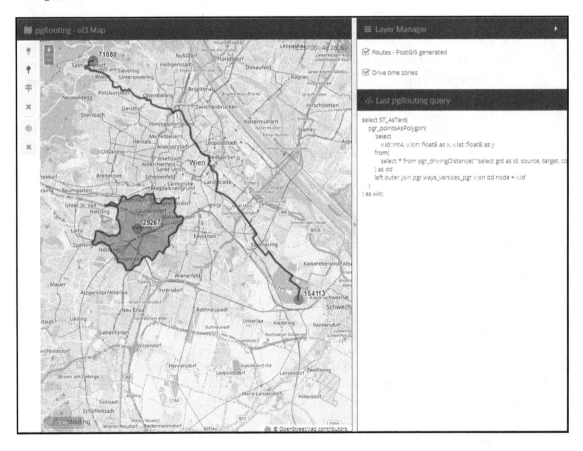

In order to feed our web app, we need to prepare a web service first. We have gone through creating a nice REST-like API for our WebGIS examples in the previous chapter, so this time all the maintenance stuff is going to be omitted.

At this stage, I assume our barebones web server is up and running, so we just need to plug in some functionality.

In order to perform any routing related logic, we should have the IDs of the vertices we would like to use in our analysis. Let's start with a function that snaps the clicked Lon/Lat to the nearest vertex in our network:

```
router.route('/snaptonetwork').get((req, res) => {

    //init client with the appropriate conn details
    let client = new pg.Client(dbCredentials);

    client.connect((err) => {
        if(err){
            sendErrorResponse(res, 'Error connecting to the database: ' +
err.message);
            return;
        }

        let query =
`SELECT
    id, lon, lat
FROM
    pgr.ways_vertices_pgr
ORDER BY
    ST_Distance(
        ST_GeomFromText('POINT(' || $1 || ' ' || $2 ||' )',4326),
        the_geom
    )
LIMIT 1;`;

        //once connected we can now interact with a db
        client.query(query, [req.query.lon, req.query.lat], (err, result)
=>{

            //close the connection when done
            client.end();

            if(err){
                sendErrorResponse(res, 'Error snapping node: ' +
err.message);
                return;
            }
```

```
            res.statusCode = 200;

            res.json({
                query: query,
                node: result.rows[0]
            });
        });
    });
});
```

The preceding method takes in a pair of coordinates expressing latitude and longitude and snaps them to the nearest vertex found in the network. It returns the ID of the node with its coordinates, as well as the query executed, so it is possible to see how our database is queried.

This API is used every time a map is clicked, in order to provide an input point for further processing.

In the map client, one can use buttons with green and red pins to define start and end route points, respectively. Once this is done, a button with the map directions sign symbol triggers another API method call - the one that calculates the route. It takes in the IDs of nodes to calculate the shortest path between them:

```
let query =
`select
    ST_AsText(
        ST_LineMerge(
            ST_Union(ways.the_geom)
        )
    ) as wkt
from
    (
        select
            *
        from
            pgr_dijkstra(
                'select gid as id, source, target, length_m as cost from
pgr.ways',
                $1::int4, $2::int4
            )
    ) as route
    left outer join pgr.ways ways on ways.gid = route.edge;`;
```

Since most of our API methods body is pretty much the same, only the actual SQL executed is presented.

We have seen a similar query already, when explaining how to get from a list of coordinates and edges output by the Dijkstra algorithm to the actual geometries representing the edges.

The preceding version does a few more operations - it unions the separate LineString edges into a MultiLineString geometry, then joins the MultiLineString into a single LineString to finally encode it as a WKT geometry, so our client app can handle it.

Because we use the minimal signature of the Dijkstra algorithm, it assumes the graph is directed, and this can be seen when our start point snaps to the edge of a dual carriage way and the general direction of a route is backwards:

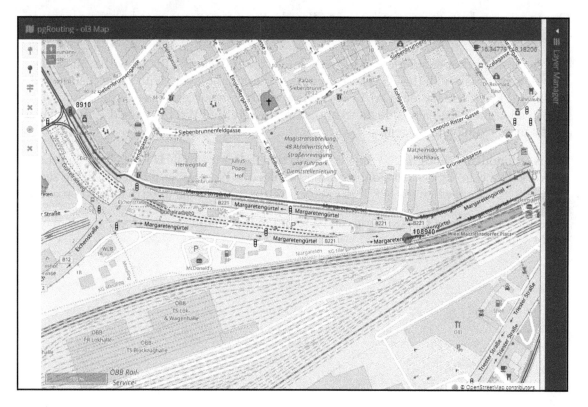

We have a final method to implement in our pgRouting API - we need to calculate drive time zones. In this scenario, we take in a node ID and the driving time in seconds and output a polygon representing an alpha shape:

```
let query =
`select ST_AsText(
    pgr_pointsAsPolygon(
        'select
            v.id::int4, v.lon::float8 as x, v.lat::float8 as y
        from(
            select * from pgr_drivingDistance(''''select gid as id, source,
target, cost_s as cost from pgr.ways'''', ' || $1 || ',' || $2 || ')
        ) as dd
        left outer join pgr.ways_vertices_pgr v on dd.node = v.id'
    )
) as wkt;`;
```

Once again, we have seen a similar query already; the difference is encoding the geometry as wkt.

The client side is rather simple and is pretty much about declaring the UI with some basic interactions, a map with the OSM base layer, and two vector layers with some customized styling. API calls use standard AJAX requests and then, on success, the executed query is displayed in the right-hand side panel and, also, the returned WKT is parsed and displayed on the map. Nothing fancy really.

It is worth noticing though, that the map projection in this example is EPSG:3857 and the network data is in EPSG:4326. So the geometries need to be re-projected before being sent out to the API and, when the WKT geometry is retrieved, it needs to be displayed on a map.

The full source code of this example can be found in this chapter's resources, so you may study it in detail as needed.

Summary

Routing algorithms may be used with some more imagination that the actual road related routing - it is just a matter of defining a specific meaning of a cost of traveling via an edge. With the appropriate data, one can build routing solutions for hiking paths, calculate routes that avoid built-up areas, or take into account some road works that add penalty costs to some edges. It is also possible to calculate drive time zones, and from there, one could go on to defining the best locations of service centers.

As we saw, using pgRouting is rather straightforward and its usage is usually down to `select * from <routing_algorythm> (<SQL for edges>, start, end, <options>)`. This makes it very easy to start with and then, as one becomes more familiar with the available functionality, to tweak the function parameters in order to improve the achieved results.

Exposing the functionality of pgRouting via web services is also quite simple, and from there, we're just a step away from consuming such services by external clients.

To summarize this into two words: pgRouting rocks (as well as PostGIS of course)!

As a final note it is worth remembering that, for larger graphs, it is good practice to work on a subset of the data for better performance - one could pre-filter the data with a bounding box to limit the number of edges to work with.

Index